W9-BYE-312

Also by Michael Novacek

Dinosaurs of the Flaming Cliffs

Time Traveler

Time Traveler

■ ■ ■

IN SEARCH OF DINOSAURS AND ANCIENT MAMMALS

FROM MONTANA TO MONGOLIA

Michael Novacek

FARRAR, STRAUS AND GIROUX · NEW YORK

Farrar, Straus and Giroux
19 Union Square West, New York 10003

Copyright © 2002 by Michael Novacek
All rights reserved
Distributed in Canada by Douglas & McIntyre Ltd.
Printed in the United States of America
First edition, 2002

Library of Congress Cataloging-in-Publication Data
Novacek, Michael J.
    Time traveler : in search of dinosaurs and ancient mammals from Montana to Mongolia
/ Michael Novacek.
        p.   cm.
    Includes bibliographical references and index.
    ISBN 0-374-27880-6 (hardcover : alk. paper)
    1. Novacek, Michael J.   2. Paleontology—Field work.   3. Paleontologists—United States—
Biography.   I. Title.

QE707.N68 A3 2002
560'.92—dc21
[B]

2001040438

Designed by Debbie Glasserman

www.fsgbooks.com

1   3   5   7   9   10   8   6   4   2

*To my teachers—Peter Vaughn, Jay Lillegraven, Don Savage,
Bill Clemens, and, most important, my mother and father*

One cannot choose but wonder. Will he ever return? It may be that he swept back into the past, and fell among the blood-drinking, hairy savages of the Age of Unpolished Stone; into the abysses of the Cretaceous Sea; or among the grotesque saurians, the huge reptilian brutes of the Jurassic times. He may even now—if I may use the phrase—be wandering on some plesiosaurus-haunted Oolitic coral reef, or beside the lonely saline lakes of the Triassic Age. Or did he go forward, into one of the nearer ages, in which men are still men, but with the riddles of our own time answered and its wearisome problems solved?

—H. G. Wells, *The Time Machine* (1895), Epilogue

Contents

■ ■ ■

Time Traveler

I

. . .

Dinosaur Dreamer

As odd as it may seem, Los Angeles is a particularly good place to become a paleontologist. I didn't always appreciate this. Indeed, as a young boy I sometimes resented deeply an urban incubation that kept me from steady contact with nature. On rare days there were opportunities. When a hair-dryer wind from the desert blew the smog offshore, I would lope up to a hilltop street appropriately named Grandview. From that high place I could see the whole sweep of the basin, from the slate surface of the Pacific to the sinuous line of the San Gabriel Mountains. Without this cleansing wind, however, the view was depressingly predictable—a bristle of telephone poles and TV antennas that looked like charred trees rising in the haze.

As an everyday experience in Los Angeles, real nature was something you saw on television. Cathode-tube cowboys raced against a backdrop of nature—always the same nature, a mass of sandstone splintered into rocky slabs like planks on the deck of a scuttled ship. Many years later, on trips with college mates to the Mojave Desert, I learned that this location for TV and movie shoots, Vasquez Rocks, had been degraded to a roadside rest area off the freeway. Power lines, airplanes, and streams of cars could no longer be reliably excluded from the frame. As I explored this thoroughly humanized terrain, littered with cigarette butts and disposable diapers, I contemplated sadly my childhood fantasy that Vasquez Rocks was a place of mystery, maybe even a place where dinosaur bones were buried.

In those early years, I hardly seemed cut out for bone hunting anyway. I was a skinny, spindly-armed boy, a kid with freckles, thick reddish brown hair, and a cowlick. The kind of kid who would stand earnestly by his lemonade stand on the sidewalk, more of a Beaver Cleaver type than a young Indiana Jones. I was not much of an athlete, and my nearsightedness required gigantic tortoiseshell glasses by the third grade. But I could run fast and far, climb pretty well, and balance decently on a narrow fence. I was not afraid of heights, but I hated closed-in spaces. I was not at all reluctant to be alone over stretches of time. There were other tendencies that prepared me for the life of a paleontologist. I liked crawling around in the dirt and mud, turning over rocks, and looking at things through binoculars and microscopes. I liked books, ones on dinosaurs of course, but books on just about anything—chemistry, astronomy, mountaineering, biographies, fiction, Mark Twain, Jules Verne, H. G. Wells, Robert Louis Stevenson, Charles Dickens, Alexander Dumas, Superman, Batman.

When I was ten, I was not quite old enough for the parental ban to be lifted on viewing the classic 1933 film *King Kong*. After all, it was rather improper for a youth to witness a hundred-ton gorilla carry a young woman in shredded clothing around in his hairy hand. But one night, after endless and annoying entreaty, I got special permission to watch *King Kong* on television. Mom, who was watching with me, apparently dozed off, and so did I. I heard a dull thud when my dad set his guitar case down on the burnt orange carpet. He had a late-night music job. It must have been at least two in the morning. At my father's command, I staggered off to the small bedroom where my brother and I shared bunk beds. I crawled up to the top bunk and lay there thinking about the movie. I tried to reconstruct my favorite scene—King Kong meets *Tyrannosaurus rex* near a giant log wedged between the edges of a deep, fern-stuffed chasm in the jungle. I replayed the scene in my mind several times, with variations. King Kong, as per the movie, vanquishes *Tyrannosaurus*. King Kong and *Tyrannosaurus* are entangled in a death plunge to the canyon floor. King Kong and the dinosaur, exhausted, drift away from each other to their respective corners of the ring. Even more gory PG-13 variations came to mind—the tyrant king disembowels the gorilla with a single slash of its scimitar teeth . . .

I could hear Dad laughing and softly talking to Mom. "Shirl, you should've heard that bass player tonight. Wow, he could play everything." Dad played jazz and those classic tunes by Rodgers and Hart, Billy Strayhorn, and others. He could also sing them spectacularly well. Only in later years did I really come to appreciate the singer and the songs. On this night, as on many others, I pulled out my transistor radio from the cleft where the bunk met the wall. The radio had a brittle plastic shell and some gouge marks for its ill-fitting screws that I'd made when I struggled to build the radio from a kit one Christmas Day. In the soft light from outside my bedroom, I could barely make out the words "Made In Japan." I turned the knob of the crude analog tuner as the radio popped and whined like the control panel in Flash Gordon's rocket ship. To my great delight, I captured the weak signal of Elvis's "Jailhouse Rock."

That night I had a variation of a recurring dream. I was walking on a red sandy hill, swinging a rock pick. An army canteen hung from my belt. I wore a wide-brimmed hat, like those of scientist-explorers in books on paleontology or even in *King Kong*. As I climbed the hill, a magnificent skeleton of a familiar carnivorous dinosaur stretched out on the ground below me. The skeleton was not completely exposed. I could see the tip of the snout and the rapiers of its front teeth reflecting in the sun. Amazingly, a couple of feet away, the brow of the eye socket—part of the same skull—was poking through the sand. Farther down were the small claws of a ridiculously tiny forelimb. A few other parts of the skeleton were exposed—a shoulder blade, some vertebrae, the stout upper leg bone, and a giant claw on the three-toed hind foot. The tail vertebrae extended a long way on the top of the hill, burrowed into the ground like a sand snake, and then emerged seven feet farther down the slope, the few tiny bones at the tip of the tail not much bigger than some of the wooden spools in my Tinkertoy box. It was all there. Every bone, process, and articulation was spread out before me, from the small openings of the nasal passages to the tip of the caudal vertebrae (like other intense dinophiles of ten, I knew many of the technical terms for dinosaur bones). I knelt down near the skull and started carefully cleaning the surface of the bone with a soft, broad paintbrush, like those in the turpentine can in our garage. As I worked, I felt alone in the world. The desert was motionless, devoid of living, breathing things, only a place where a line of red sand met the sky. There was no wind, only the low hum of the engine that pumps blood through your inner ear. Soon, though, I began to feel the intrusion of a shadow at my back. I turned around and was transfixed as a giant cumulus cloud coalesced into the form of *Tyrannosaurus rex*.

Tyrannosaurus

At the time I first learned about them, dinosaurs were believed to have been sluggish, dumb reptilian creatures. Their relationships with living groups were vague, though crocodiles and lizards at least provided some inspiration for what dinosaurs must have been like. Indeed, many science-fiction filmmakers recruited lizards and attached horns or spikes to them when they wanted live-action dinosaurs. They then filmed the embellished lizards attacking other lizards or even cavemen. Even as a kid I knew the juxtaposition of these faux dinosaurs with humans was absurd—humans after all appeared on the earth 60 million years after the big dinos went extinct. But in the late 1950s, neither I nor paleontological experts knew many things that we know today. For one thing, we didn't know that some dinosaurs indeed survived that massive extinction event at the end of the Cretaceous period, the last chapter in the age of the dinosaurs. The survivors were of course birds. We have strong evidence that birds likely evolved from a subgroup of dinosaurs including active, predaceous, and probably very intelligent forms like *Tyrannosaurus* and *Velociraptor*. Today a few scientists still object to this connection between birds and dinosaurs, but they do so in denial of a mass of accumulating paleontological data. This includes newly discovered fossils from China that show that nonflying dinosaurs much like *Velociraptor* had feathers!

As with many other young lives, mine was devoted to desperately re-creating—through dreams and mini-expeditions—a landscape that erased the banal ambience of my "hometown." When we think of nature, we may think of a magnificent African savanna populated with lions, giraffes, and herds of elephants. But what is nature to a kid in Los Angeles? To me it was a rather limited list of creatures: an overturned potato bug with its half-dozen legs desperately flailing to find hard ground, an alligator lizard popping out when lifted by its shedding tail, some fat tadpoles in a garden pond, a tiger swallowtail butterfly flickering yellow and black in the geraniums, a pill bug, an ant colony, a rat in a palm tree.

A neighborhood encounter with dinosaur bones or other fossils was of course essentially impossible. This experience would require some imagination. For me, a vacant, weed-choked lot became a fossil-strewn desert with endless prospects. One such area was a spit of eroded sand and gravel that divided a broad street a few blocks from my house. Crossing this escarpment one day I saw a string of fragmented concrete blocks that looked for all the world like the vertebral column of a long-necked sauropod dinosaur. I recruited a group of friends, including my rather forcefully conscripted younger brother, to excavate this monster. Our make-believe work on "Dinosaur Island" was my first field expedition.

I read about real field expeditions, especially ones that brought bone hunters to the canyonlands of Montana and Wyoming or the empty, windswept deserts of Mongolia in Central Asia. I knew with satisfaction that there were even great excavations and discoveries within the city limits. The very center of the L.A. Basin—its lowest part—is a jumble of upscale department stores and high-rise apartments along Wilshire Boulevard's "Miracle Mile." The only nature in this agglomeration is a poorly manicured park of tawny grass and dusky oak trees, where in places small pools with an unctuous film of slow-moving water covers the stews of bubbling tar. The tar gives the park its Spanish name, La Brea (although the sticky substance in the pools is actually asphalt, a mixture of different forms of tar and sand). Thousands of years ago, huge numbers of mammals, birds, lizards, frogs, and other creatures congregated around these pools for water. A great many of them fell in and died trying to struggle out. This accumulated carnage makes for one of the world's richest and most remarkable fossil sites.

I went to Rancho La Brea Park as often as possible with my parents to watch history in the making. I stared at those pools, waiting for some hapless squirrel or sparrow to go too close and get stuck fast. It takes many hours or even days to witness such tragedies—indeed I don't think I ever saw any—but the evidence for mortality was abundant. It was a thrill to see a feather sticking out of the tar like a quill pen in an inkwell, or the leg bone of a rabbit picked clean by a predatory bird all the way to the tuft of dandelionlike hair around the tough, inedible foot. Death did not seem in any way dark or frightening; it seemed an inevitable conclusion whose circumstances left fascinating evidence of an earlier life. I even made a fleeting discovery that there was a La Brea in my own neighborhood. Romping through a broad swamp destined to be

A detail from the mural of the La Brea Tar Pits 12,000 years ago,
by Charles R. Knight, without the horses and camels

reclaimed as a golf course, my friends and I spotted some black, vis-
cous ooze with that familiar rainbow sheen lazily swirling around
the stalks of high grass and bamboo. "It's a tar pit!" we shouted.
Moments later we were bitterly disappointed to find nearby a pile
of ulcerating oil drums.

My favorite features at Rancho La Brea Park were some life-
sized statues of saber-toothed tigers and imperial mammoths. These
creatures of course no longer lived in the L.A. Basin and their
bones were not visible on the surface of any of the pools. However,
a stupendous array of tar-stained black bones of such animals filled
the halls of the Los Angeles County Natural History Museum,
some ten miles away. These were the remnants of an ancient and
populous community of big mammals. They were an odd bunch, a
combination of the bizarre with the familiar—cat skeletons as big
as bear skeletons, and elephantine skulls that looked like helmets of
ancient Greek warriors, only with tusks that extended so far they
curled back on themselves and crossed each other.

Above these mounts was a mural by the great painter of prehistory, Charles R. Knight. This is La Brea 12,000 years ago. The setting is all southern California. The stage is a burnished field of
grass, the kind that becomes a tinderbox in the dry canyons of the
Santa Monica Mountains on those hot September days. A few
writhing, leafless oaks are rooted in the grass around the black tar
pools. The backdrop is the distant deep blue of a mountain range,
the same profile of peaks I could see from Grandview Boulevard—

after all, mountains don't appreciably change in 12,000 years unless a volcano punctures them or an earthquake breaks them down. Perched on a surly gray oak are a gang of crook-necked giant vultures peering down on their mates feeding on a bulky carcass. The vultures look odd and out of scale on that tree, as indeed they should: with wingspans of nearly thirteen feet, they are about one-fourth larger than their living and highly endangered relative, the California condor. The foreground of this panorama is dominated by two different and famous La Brea species. At left center, a saber-toothed cat in full profile is in a stalking mode, its jaws at open gape, yawning, displaying its absurdly elongate upper canines. At dead center are a cluster of zombielike giant ground sloths. The biggest of these is standing high on its hind limbs, its neck stretched, its nose up, like a grizzly bear sniffing the air for intruders. Flanking this center-stage group is a cohort of dire wolves; their heads poke up through the tall grass ready to pounce and to dine. To the right, a bit farther back, a herd of camels stroll slowly into view, humpless, short-necked, barrel-chested, pigeon-toed. Behind the camels some stocky black-legged horses kick up the dust. As if not to overshadow this bestiary, the huge imperial mammoths, some fifteen feet high at the shoulder, can be seen in the far distance in a clearing among the oaks and below the blue mountains.

They are all dead, gone, extinct. There are no more imperial mammoths, dire wolves, La Brea camels, saber-toothed cats, and giant ground sloths. Not even those horses linger on. Horses native to North America went extinct around La Brea time; the horses that ran wild in the West or were mounted by Indians were import models, brought in by Spanish conquistadors and other explorers. Charles Knight's vision of life in Los Angeles 12,000 years ago was grander than the African savanna. I was struck by the fact that Los Angeles a few thousand years ago had bigger, more awesome animals than any place on earth today.

The La Brea Tar Pits and the fossils embedded in them were, to me, a miraculous gift to Los Angeles. In Catholic school, a nun who had visited Lourdes regaled us with tales of the astounding cures that came to pilgrims who partook of the holy waters there. But Lourdes did not seem at all interesting to me. Even at the age

of ten, I had developed a skepticism for the mystical and the unex-
plainable. La Brea was both fantastic and *real*; religious lore could
not compete with its vivid evidence. My predilection for empiri-
cism was not always rewarded. The same nun who waxed poetic
about Lourdes had divided our fourth-grade class into three sections according to a scale of perceived intelligence and acceptabil-
ity. The west two rows were the "Dummies," the east two rows, the
"Smarties," and the middle purgatorial row, wherein I held a place
of dubious honor, the "Spacemen." I tried to transport myself out
of that miserable place as often as possible, usually by concealing
my favorite science book behind a catechism for a surreptitious
read. One day I savored the diorama painting of Ice Age mammals
in the large, not so easily concealed Time-Life book *The World We
Live In*. The nun swooped down on me and rapped my hands.
"There you are once again reading about monsters!" she shouted as
she threw the book to the floor and gave it a permanent stamp
with her dainty little shoe. I was held after school, but I got the
book back. Later that afternoon I sat on the porch at home reading
the forbidden *World We Live In* and sipping lemonade. I day-
dreamed about a Los Angeles lost in time with its ground sloths,
mammoths, and saber-toothed cats. I extended the dreamscape beyond my hometown to the Montana badlands where *Tyrannosaurus*
took rest and to the empty red deserts of Central Asia where the
first dinosaur nests, and the first complete skeletons of primitive
shrewlike mammals (presumably the marauders of dinosaur nests),
had been discovered. I dreamed of all the continents, canyons, and
quarries still waiting to be plundered for fossils. It was good to be a
"Spaceman."

2

. . .

The Canyon of Time

My parents were well aware of my tensions with the nuns over books about lost worlds and prehistoric monsters. Yet aside from some expected corrections to unruly behavior, they greatly encouraged my curiosity. My dad is a gentle man whose passion as a jazz guitarist and singer seems itself a force of nature. His huge nose made him more handsome, as if it were needed to damp the corona of his blue eyes and a smile that fanned out like a horizon. In those days he had sandy curls above his high forehead, a feature accentuated in a cartoon that captures his essence well. The picture, done by an artist friend many decades ago, depicts Dad playing a left-handed guitar and wearing a corny Western plaid shirt with pearl buttons, a bandanna, and cowboy boots with spurs. "That's you, Al, you to a tee," everyone says, even though the picture title calls him "Lefty" and Dad is a right-handed guitar player and is more likely to be in black tie than boots and spurs. The picture still hangs in my parents' living room, surrounded by little shrines to their children and grandchildren.

No portrait of my mother hangs in the house, but she deserves one. Like her mother and her sisters, she is a beauty. There is a family treasure, a studio photo taken when she was in her upperclassman years at UCLA at the end of World War II. The planes of her face, with its sharp delicate nose and perfectly sculpted cheekbones, rival the architecture of Joan Fontaine's or other cinema love goddesses of the 1940s. She had a body to match. A photo of her from the same era shows her holding a rifle, wearing a short dress, and sporting tortoiseshell sunglasses that look very Prada, with one high heel on the bumper of a Kaiser sedan. One friend said when he saw the photo, "Marilyn Monroe, with longer legs." At her high school in Chicago she was named Miss Sweater Girl. She first met my dad when she came back to Chicago from California to see her parents. It was just after World War II and jazz, Sinatra, and zinc bars bristling with martinis were contributing to one continuous victory party. Dad was playing with a trio at the posh Brown Derby. Mom went there with her alluring girlfriends, saw this guy playing guitar on stage, and announced that she would marry him. She instantly landed a job as a cocktail waitress at the establishment to increase the probability of capture. Dad didn't have a chance. When I was six or seven and I went with Mom to the supermarket, sinister, green-skinned men would sometimes come up to her in the parking lot and say something in low tones that I couldn't make out. She'd fire back, "Back off, buster, you see this ring on my finger. Now get away from my son and me."

Mom's work as a cocktail waitress was only a transient and strategic avocation. Her lifelong career has been devoted mainly to managing her four sons, but she has demonstrated extraordinary energies in writing, horticulture, and certainly science. After her fourth child was born, she went back to college, started taking biology and botany courses, and outshone all her much younger academic peers. That inherent curiosity that is so scientific had always been a characteristic. When I picked up the translucent sheath of scales of a lizard's shed skin, she would exclaim, "Isn't that interesting!" She meant it. And I accepted that kind of attention as a natu-

Turritella, a fossil snail from the Santa Monica Mountains

ral parental response. As the years passed, I came to recognize my fortunate situation. My mother was a patron, protecting me from the prosaic intrusions of school and daily routine so that I could get on with my explorations and take my notes and make my drawings.

Indeed, my parents made it possible for me to discover that the Los Angeles area could offer more than museum exhibits. There were actually opportunities for real fossil hunting. Within an hour's drive into the Santa Monica Mountains road cuts of bronze sandstone contained a cornucopia of marine creatures. These looked very familiar—the rippled amorphous shells of oysters and spiraled cones of sea snails. But it did not require great expertise to tell these were fossils. They were hard in an ancient, enduring way, having taken on the hue and feel of the surrounding rock. They were indeed easily discriminated, not by their shapes but by their qualities of preservation, from shells washing up on the current shoreline a few miles to the west 15 million years later. The Santa Monica Mountains were simply piles of sand laid down when the Pacific invaded and retreated from the land over millions of years during a time called the Miocene epoch. The amount of rock accumulated during this repeated cycle of flooding and retreat is staggering; in places, the sediments are nearly 30,000 feet thick, as thick as Mount Everest is high. Most of these layers are covered with dusty shrubbery known to ecologists as Mediterranean zone chaparral, a carpet that effectively conceals the rocks and their fossils. But the irony of human intrusion is that it sometimes reveals a natural history that we otherwise might never know. Scattered road cuts and housing tracts offered the only chance to dig into the rocks for fossils. We were combing ancient beaches for the animals that had peppered them with shells.

Those shells and those Santa Monica Mountains were old, recording a calendar ticked off in millions of years rather than the measly thousands of years evoked in the bones of La Brea. But an encounter with an even more remote timescape and a vastly more enduring column of rock required an escape from L.A. On rare occasions we piled into a two-toned Chevrolet station wagon and headed for places with big rocks—Yosemite, Mount Whitney, and especially the Grand Canyon. En route we passed through more rocky terrain. We bisected the Mojave Desert at night across a pockmarked landscape, bleached and ashen in the light of a full moon. The shadowy hills on the horizon were the huge surf of a black ocean. On this nocturnal drive I was often the only other one awake besides my dad. I rode shotgun, pretending to change the gearshift as Dad took on a winding upgrade. Dad had commuted many nights across this desert to play music in Las Vegas. He told me of the times when Red, the bandleader, would have a fight with the management of the Sands or the Thunderbird Hotel. Red would buy a fifth of scotch and say, "C'mon, boys!" Hopping into a long white Cadillac convertible, they'd scream back to L.A. at eighty miles an hour. "We didn't have much time to take in the scenery then," Dad said. At that moment, our headlights caught the fiery reflection of the eye of a rattlesnake warming itself on the blacktop.

By dawn we were moving up the long slopes to the Coconino Plateau, where the landscape was obscured in shadow and the earth seemed transformed into something cooler and more refreshed. For the first time I felt that acrid pinch of the nose from smog or dusty air disappear. The aroma of the dew-coated sage mixed with that of juniper. I could smell the ice-cold martinis on my parent's glass coffee table.

The flat highlands covered with its thin veneer of vegetation were soon violated by an outrageous gash in the earth, the Grand

Canyon. A guitarist, his wife, and three small boys edged out to the rim overlooking the Bright Angel Trail. In those days, the scenic overlooks were not festooned with concrete walkways and safety railings. My mother always made the same corny joke, "Watch that first step, it's a lulu!" We contemplated with a bit of vertigo and morbid fascination what it would be like to plummet some 5,000 feet down, ricocheting off the candy-striped cliffs to plop like a pebble into the ribbon of the Colorado River far below.

Even before I ever saw the Grand Canyon I knew that such a fall off its rim also meant an unparalleled descent through earth time. Since then I have learned that the canyon itself was born about 20 million years ago, when the Colorado River began scouring out the rising land. The course of the river, as it carved through the rocks, shifted through time. It has been proposed by some geologists that the ancestral Colorado actually took a big arc and flowed back northward through what is now Utah and then veered west through present-day Nevada and California, eventually emptying into the Pacific Ocean. Sometime around five million years ago it changed course southward and exited the canyon via its present route, its meandering course marking the jagged boundary between southern California and Arizona. Here the Colorado's sluggish waters hardly seem able to move a grain of sand. The river's power has been diminished as its waters are siphoned off to irrigate fields of lettuce and watermelon. What is a terrifying torrent in Lava Falls and other rapids of the Grand Canyon becomes a trickle as the river meets the Pacific at the uppermost reaches of the Gulf of California.

But this tortuous history of the Colorado River is relatively brief; it merely chronicles an end-zone event, an incision on a vast, multilayered monument of rock that has been built up for more than 1.7 billion years. The canyon cliffs are bands of massive rock stacked on top of each other. They were wrought by big forces— raging rivers, volcanic flows, flooding seas. There are even ancient coral reefs turned to stone, built from the mighty cooperative labors of millions of marine creatures, leaving behind their infinite maze of shells and skeletons of calcium carbonate. Most of the bands of rock are in fact products of the sea, laid down as the bot-

A panorama of the Grand Canyon from J. W. Powell's 1895 book, *Canyons of the Colorado*

tom layers of deep oceans or shallow coastal waters. The base of a mountain range about 1.7 billion years old forms the canyon bottom. These are contorted rocks squeezed by eons of heat and pressure as the earth's crust buckled with earthquakes and mountain building. Above these rocks is a mysterious gap of about 450 million years. What happened here? The rocks were likely eroded away in a highly unstable landscape, where torrential rivers or other forces scooped out big chunks of crust. Lying directly on top of the canyon floor is a layer of rock called the Bass Formation, which is 1.25 billion years old. It contains some of the earth's oldest organisms—primitive stacks of algae called stromatolites, whose relatives still survive today in some isolated refuges such as coves and shallow beaches on the Australian coast. About a billion years ago, the lower rocks of the canyon were formed, ancient layers of mud from retreating seas and even a belt of volcanic lava called the Cardenas Basalts.

Give or take a few more layers and a few more hundred-million-year gaps, and then the first stately band of vertical cliffs is exposed. This is the Tapeats Sandstone, which was laid down about 550 mil-

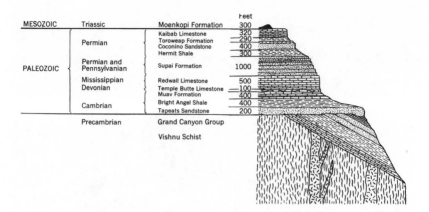

			Feet
MESOZOIC	Triassic	Moenkopi Formation	300
PALEOZOIC	Permian	Kaibab Limestone	320
		Toroweap Formation	290
		Coconino Sandstone	400
		Hermit Shale	300
	Permian and Pennsylvanian	Supai Formation	1000
	Mississippian	Redwall Limestone	500
	Devonian	Temple Butte Limestone	100
		Muav Formation	400
	Cambrian	Bright Angel Shale	400
		Tapeats Sandstone	200
	Precambrian	Grand Canyon Group	
		Vishnu Schist	

A geologic section of the Grand Canyon

lion years ago, a time when one of the early radiations of more complex plants and animals, creatures with segments, branches, legs, even eyes, are preserved in the rock. Capping the Tapeats Sandstone are the more gentle, dusky slopes of the Tonto Plateau. Above this plateau is a steeper, but not vertical band, the Bright Angel Shale, formed when a great ocean was again advancing. Above this layer is the series of seven named layers of picture-postcard rocks, which most visitors associate with the chromatic Grand Canyon. My favorites were always the Redwall Limestone, a blank rampart at about the midsection that represents a 330-million-year-old sea bottom, and the Coconino Sandstone, a white cliff a few layers below the rim of the canyon, which is swirled with beautiful ordered streaks called crossbeds. These crossbeds mark the surfaces of ancient sand dunes marching southward with the prevailing wind 270 million years ago, when they were deposited.

What perplexed me the most in my boyhood visits to the Grand Canyon were its topmost layers. The rim is formed by the Kaibab Limestone, another ancient sea bottom stuffed with fossils that is about 250 million years old. Where are the younger rocks? It seems a huge rock record extending back a quarter of a billion years has been erased by erosion in the near reaches of the canyon. Such younger rocks are preserved to the north in places like Zion National Park and Bryce Canyon. So the Grand Canyon provides a slice of one of the most enduring and varied chunks of the earth's crust, but it does not bracket a complete rock record of the earth. There are ancient rocks in Greenland and Australia that are older than 3 billion years, and studies of ancient meteorites that have fallen to the earth indicate that that our planet and our solar system are at least 4.5 billion years old. At the other end of the calendar, there are many rock layers, including beds that preserve dinosaurs and ancient mammals, that represent "merely" millions or thousands of years. This rather constrained time span of the Grand Canyon is an important fact. Most of its millions of visitors are surprised to learn (if they indeed learn at all) that its youngest rocks preserve a world that existed about 100 million years before the appearance of the first dinosaurs.

Taking account of this entire timescale, this entire 4.5-billion-year history of the earth, has been the business of geologists for a few centuries. Even Leonardo da Vinci, in his codex of the late fifteenth century, recorded layers of ancient sea rocks containing fossils in the mountains of northern Italy. Da Vinci noted that layers bearing sea creatures were sandwiched between rocks that looked as if deposited on land, thus recording a cycle of invasion and regression of the seas. Leonardo would probably have a reputation and a hold on geology much as he has on art, except that, as noted by many historians, he rarely finished any project, however brilliant, that he started. Later, Nicholas Steno refined Leonardo's perspective and developed perhaps the first version of what we call stratigraphy—the science of strata, or layers of rock. Steno made the utterly simple but emancipating observation that younger rocks almost always lie on top of older rocks. Others, like the nineteenth-century British geologist Charles Lyell, made exhaustive charts of rock layers in Europe, noting their age as deduced from the marine shells and other fossils contained within them. Lyell and his successors thought big when it came to time. They suggested that the rocks they were studying had been laid down over millions of years. It was Lyell who planted the appreciation of this vast timescape in Charles Darwin. A key book read by Darwin during his voyage around the world on the H.M.S. *Beagle* was Lyell's *Principles of Geology.* Thus Lyell's concept of an earth refashioned over millions of years by alterations to the continents and ocean basins

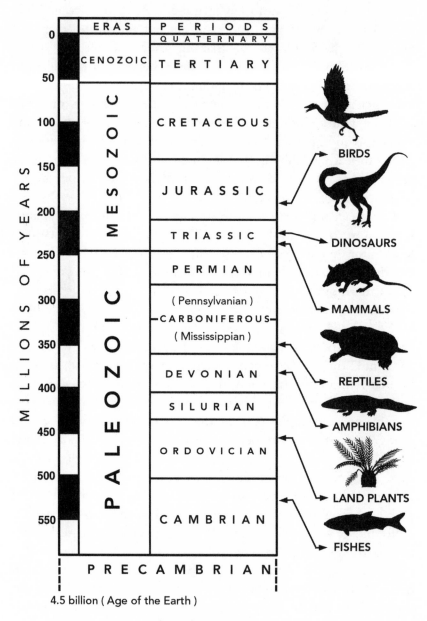

The geologic timescale

provided the dynamic and enduring stage for living creatures. It made plausible Darwin's idea of evolution of species through natural selection—a process that may require millions of years.

Since the beginning of the twentieth century, this timescale has been refined and notably expanded. Not only fossils are useful for setting the rocks' age. Tiny traces of radioactive elements in the rock transform their structure like the tick-tock of a well-made clock. Charged atoms (ions) of uranium, potassium, lead, carbon, or other elements lose or gain subatomic particles at a constant rate; knowing this rate and the percentage of transformed element in a given rock sample, one can determine when the rock was first solidified and formed. This age estimate is a complicated calculation that requires measuring the contamination or leakage of the product ion, but the basic principle applies to all dating of rocks that uses radioactive elements.

A combination of radiometric dating and careful comparison of fossils yields a very detailed worldwide geologic calendar. Modern geologists have achieved a major refinement of the subdivisions of earth time proposed by Lyell and other earlier scholars. Nonetheless, the first four billion years of that calendar are still very crudely resolved because that vast expanse of time is so poorly represented in rocks and fossils. We call this first dark chapter the Precambrian Era. Then, from about 570 million years ago (about the age of that Tapeats Sandstone exposed in the lower reaches of the Grand Canyon) until the present day, the record is spectacularly enriched with the fossilized remains of extinct life. The first of these better-preserved chapters, the Paleozoic Era, is dominated by marine life, although it also records the dramatic emergence onto land of plants, animals, and other organisms. The second of these subdivisions, the Mesozoic Era, starts at about 250 million years ago, and is commonly if somewhat misleadingly called the age of the dinosaurs. (A lot of other important things happened during the Mesozoic, including the rise of the flowering plants and the pollinating insects, such as wasps and bees, and of course our own group, the mammals.) The third great subdivision, the Cenozoic Era, begins roughly 65 million years ago, when all the dinosaurs went extinct except for one specialized group of flying, feathered dinosaurs, the birds. This phase experienced a dramatic rise of mammals, lizards, snakes, and other land creatures, as well as big changes in the marine realm. Although we sometimes formally distinguish the 15,000 years or so since the last great ice age as a time of man, we would better regard it as simply a footnote to the Cenozoic. After all, the human lineage, represented by *Australopithicus* from Africa, has only been around for the last 4.4 million years and the modern human species, *Homo sapiens*, only for a few hundred thousand years—less than 1 percent of the time since the extinction of the nonbird dinosaurs.

When I was a boy, I took many of these statistics to heart, eagerly memorizing the dates and the ages. It is impossible, however, fully to appreciate the scope of time measured in thousands, millions, and certainly billions of years. Remote time, or deep time, may be beyond our ken. Likewise, the awareness that there are hundreds of millions of galaxies, each with hundreds of millions of stars, in the known universe leaves us awed but little enlightened about the true nature of the universe.

Maybe that's why the Grand Canyon as a metaphor for the immensity of time is so compelling. Its scale is huge but not beyond our tactile appreciation. One can even walk from the top to the bottom of its timescape in a single day. And even though the canyon does not record all of earth time, the 1.7 billion years it does preserve is time enough. The mountains at the beginning of this history were replaced by coastal seas, then mudflats, a varied coastline, a major volcanic eruption and lava flow, more coastlines and shallow seas, deeper oceans, shallower oceans, to deeper oceans again, emergent land, a dense forest of ferns and cycads, a vast sand desert, an undulating coastline with rising and falling seas, and, at the topmost Kaibab layer, another deep ocean bottom. We are lucky to have a fossil record for at least part of the history of life. Thus one can do better than simply fantasize about the evolutionary epics of animals and plants that were superposed on this changing Lyellian earthscape.

Over the years I have often hiked the Bright Angel Trail and the Kaibab Trail to inspect these layers of the Grand Canyon at close range. But I confess that during the ascent, geology is about the last thing on my mind. This was especially true during a jaunt I took with some friends at the end of May in one of those infamous canyon heat waves. Phantom Ranch, at the bottom of the canyon, registered 108 degrees Fahrenheit at five o'clock in the afternoon. In the epitome of bad timing, the traditional rim-to-rim long-distance run was scheduled for that same infernal week. The racers obviously did the easy part, the descent, first. But the withered runners arrived at the bottom of the canyon in no condition to get up the other side; they had to be hauled out on mules.

We were an equally sorry lot, I among the worst off. A cactus had punched a hole in my primary water jug, so I had some makeshift containers dangling from my vest. The same cactus had also ripped a huge, obscene hole in the crotch of my shorts. Perhaps the only advantage to this wardrobe modification was that it made for more convenient dealing with diarrhea, a common symptom of acute heat stress. These various injuries and ailments were complemented by a pattern of blisters that made my feet look like bubble pack. During our death march we encountered another group of hikers in even more desperate straits. They crouched in the meager shade of an overhang, awaiting a rescue. Two of their members were disastrously overheated, as white as the Coconino Sandstone, with trembling limbs, and vomiting a substance that looked like tapioca pudding. We combined with this group and pushed on in delirium along the Tonto Plateau back to Phantom Ranch. As we crossed the river at 3:00 a.m. in full retreat, we took bets on how fast the thermometer would rise from its centenary mark before we could escape. That hike nearly destroyed my interest in and certainly my affection for the Redwall Limestone.

At the time of my first visits to the canyon with my family, such excruciating exploration was far in the future. I only took it in from afar, from the top of the rim or on a brief skip down the first hundred yards of the Bright Angel Trail. This rather slim experience did not subdue my naive sense of familiarity. Indeed, after a couple of visits to the canyon I thought I knew the place, and I was impressed. After all, was there any more spectacular landscape on earth? Wouldn't everything else seem downscale, undramatic? You could fit an outcrop full of fossil shells from the Santa Monica

Mountains or the whole collection of tar pools from La Brea in a corner of the Grand Canyon and never even run across them. I imagined living in the canyon and becoming the one person in the world who actually had roamed every square foot of its gullies, crevices, and caves, never returning to the same spot. I was cast adrift on a great sandstone sea. I was lost in the rocks and lost in time.

3

▪ ▪ ▪

Early Expeditions

I begged my parents to return to the Grand Canyon whenever the family had any prospect of a long weekend or vacation. We plotted trips to Arizona, dragging out the road maps on the dining-room table, as if we were charting great expeditions. Then something interrupted those well-laid plans. Things were not going so well for more conventional musicians in L.A. in the late 1950s. The rock music I found so alluring was putting the squeeze on Dad and his contemporaries. Live music was being replaced by DJs and blaring sound systems. Coincidentally, there was the temptation of a good, full-time music job in more conservative resort lands back east. Some years before, Dad had worked at the Northernaire Hotel in northern Wisconsin, a white curvaceous edifice with round portals and picture windows shaded with blue awnings. It was meant to look like the *Queen Elizabeth* docked on a picturesque lake. Now the hotel owner was asking Dad to come back. Of course the winters would be cold and long and the low season would force Dad and the trio to go on the road. But my parents exclaimed that Wisconsin had all those great things we didn't have in L.A.—boating, fishing, skiing, ice-skating, horseback riding—it was a sound bite for a kid's nirvana.

I vaguely remembered a very early summer in the north woods when Dad had worked at the Northernaire before, so I was enthusiastic, but I knew this would take me away from the Grand Canyon, the Santa Monica Mountains, and the La Brea Tar Pits. I knew enough to know that Wisconsin was mainly a land of very old, boring rocks—some hundreds of millions or even billions of years old—in many places covered by a sheet of boulder-laden soil that had been pushed and crunched by glaciers during the last million years. Some of these top layers contained rather young fossils, like those from La Brea, about 10,000 or 12,000 years old, representing the last great retreat of the glaciers. Unfortunately, none of these bone beds was as spectacular as the graveyard at La Brea. Despite the prospects of clear lakes and waterskiing, I was a little apprehensive about leaving what I viewed as real fossil country. It was a sentiment that foreshadowed a basic inclination of the profession. Paleontologists don't like trees. They obscure the rocks, and if Wisconsin had anything, it had lots of trees. As we rode the Union Pacific Railroad on our way east I pressed my face to the window at the back of the Domeliner to catch the full view of the Green River meandering through a great canyon of tilted white sandstone cliffs. I wondered if I would ever see such a grand pile of rocks again.

As it turned out, this adaptation to woodlands was short-lived. We stayed in Wisconsin only three years. Long winters, Dad's prolonged absences on the road, and the need to get back to a real city, even one with heavy traffic and smog, conspired against us. Indeed I barely need mention this interlude, except that it contained an important event in my nascent career as a paleontologist. I had a second cousin from Chicago, whom we customarily called our uncle, who was an avid amateur paleontologist. He had two pretty daughters who did not at all share his enthusiasm for fossils, but he dragged them off anyway on expeditions. I was included in these sorties, the only willing member of the team. We aimed for some small stone quarries in the dense forests about fifteen miles east of the Northernaire. These hollows exposed old marine rocks, limestones some several hundred million years old wedged somewhere between the old granitics and the Ice Age scree.

We all squatted in the quarry and started splitting big pieces of limestone with some household hammers. As we chatted about Dion and the Belmonts and Corvettes, I broke open a fat brown limestone that looked like a loaf of German rye. Out came a fossil, a creature magnificently displayed, its head in full view, with large eyes covered with myriad tiny bumps. These bumps were actually lenses, I had read, like the many lenses of a fly's compound eye. The head was sculpted with small plates; the middle section of the skeleton was brown and smooth with a slight reflection—it looked like the shield of a gladiator; and behind was a series of curved rib-like spines that reminded me of the segments of the pill bugs that were such an important part of L.A. nature. The fossil was indeed a relative of pill bugs and, for that matter, of centipedes, crabs, lobsters, spiders, scorpions, insects, and other members of the earth's most dominant animal group, the arthropods. But this particular arthropod had been extinct for roughly 280 million years. It was a trilobite, the dominant arthropod of Paleozoic seas, which went

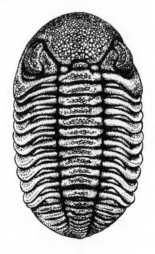

A trilobite

around scarfing up other creatures. It was a beautiful trilobite, nearly eight inches long and perfect.

My uncle grabbed it from me with trembling hands.

"Oh my God, now that's something." He made a slight sucking sound that came from deep within his chest.

"Wow!" I said.

"This must be the best thing ever from this quarry." My uncle was now standing, waving the specimen to an imaginary crowd.

I kept waiting for him to give it back. I finally got my fingers on it while he still had it in his grasp. I rubbed the smooth surface of the head and the middle section.

When we returned home my uncle told my parents that the fossil was good, so good that he should take it back to Chicago, to a scientist at the Field Museum of Natural History. Maybe they would want to include the specimen in the museum collection.

"That's great, Mike, great find!" my uncle kept saying, smiling brightly. I was pleased but a little confused. Didn't I find the fossil? What did the Field Museum need with it? They already had several thousand trilobites. My uncle wrapped the fossil in tissue paper and carried it back to the guest room. That is the last I saw of the trilobite. Maybe it is now in the Field Museum collections. I wouldn't recognize it if it were. I never asked my uncle about it after the one or two times I saw him during those Wisconsin years. He may have forgotten about it.

Despite this confusion, I only felt a muted resentment for the loss of my prize specimen. In later years, as a professional, I realized that I had more passion for collecting than for keeping fossils, as long as the specimens I found ended up in a good collection, well cared for, safe, and easy for others to study. That shouldn't be taken as high-minded generosity. It is hard to maintain a good collection responsibly, and the best collections are so because people of different skills—scientists, preparators, cataloguers, conservators—all share in their care. That is what museums are for. But I never forgot that trilobite discovery because it was the first time I felt I found something that might be exciting to others.

I was thirteen when we moved back to L.A., but by then the California prospect caused anything but exaltation. I had forged a real attachment to northern Wisconsin. I even liked trees, especially trekking through the woods in the winter following a set of deer or raccoon tracks. It was sometimes nice to get away from people, and to do it just by stepping off your front porch or heading out to the woods behind the garage at White Buck, the deserted resort lodge where we stayed during the warmer months. Besides, there was a girl, Diane; we kissed in the back of the bus on the way to Eagle River. Goodbye Diane, goodbye lake country, woods, northern lights, pristine snow, empty places. As we struggled along Route 66 with a trailer in tow I buried myself in a book, *Banner in the Sky*, by James Ramsey Ullman, a stirring story about a boy only slightly older than me who scales the Matterhorn. I wanted to trade places with that boy more than anything in the world. I was pretty taciturn for most of the drive, and my parents knew why and didn't indulge me. Then we hit Arizona and the car for the first time revved up to ascend the waves of hills rising to the Coconino Plateau. As the evening fell, cool air rose out of the canyons carrying the scent of moist juniper. I broke my vow of sullen silence.

"Can I ride shotgun, Dad?"

"Sure, but keep awake."

"How far are we going?"

"We'll drive three, maybe four hours, then we'll camp near the Grand Canyon."

"Great."

And with that little exchange I thought for the first time maybe I was never meant to stay in the green glades of Wisconsin anyway. I was back in the land of the rocks.

People often ask me about my childhood as if it forms a clear link to my present persuasions. But of course it doesn't happen that way. All that passion and curiosity for fossils, rocks, nature, all that interest in science books, could easily be swept away as the wonders of culture and society in a 1960s Los Angeles came crashing in on a teenager like one of those giant winter waves off Santa Monica Beach. Who cares about the Grand Canyon and the geologic timescale when you can put your head against a girl's leg and close your eyes, listen to the whoosh of the sea, and smell the slathered Coppertone? When I came home from the beach, barefoot, sunburnt to a crisp, long hair hiding my face, carrying the one fin I used for body surfing, my poor dad would loudly lament the metamorphosis. "Why oh why did we ever leave Wisconsin? Wear shoes sometimes. You call that rock band you play in a job, you call that music? Don't get kicked out of school like your brother did." My grandfather, who grew up in the old country of Czechoslovakia and had witnessed many political and cultural upheavals, was more succinct in his reaction. "Comes the revolution!"

Not that any of these paternal objections had much effect on my teenage soul. I proceeded to integrate myself fully into the land of the lotus-eaters along with its other young citizens. We were the new generation, the Who's "My Generation." What did Paleozoic time matter to someone who hoped that "I die before I get old"?

Then I almost got a chance to die. There was a big war in Vietnam, some ugly, hot place near China. Our high school political science teacher said that maybe there would be a million American soldiers there before long. I started college, eligible for the big universities like the University of California at Berkeley or Los Angeles, but taking the low road to Santa Monica City College in order to play music and stay out of the draft on a student deferment. I joined bands and broke up bands. We kicked out a drummer from one of them; he joined the Marines and got killed in Vietnam. We were shocked by the sudden brutality of this death and its connection with our own lives. I survived the system, though, insulated from the horror of wartime that took our poor ex-drummer and too many others.

During this early college phase, when the "So Cal" culture engulfed me, it was odd that I undertook some informal, almost subliminal training for paleontological work. A friend named Mick had a Volkswagen bus that he had set up like a desert cruiser, well fitted with water tanks, stove, refrigerator, and tarps. Mick, Russ and his brother Kevin, and my younger brother Steve (the one who'd been expelled from our Catholic high school) and I would head out to the Mojave Desert for long weekends, skipping those boring psychology classes on Monday.

These jaunts were, however, educational. A need for locating good old mine sites drove me to learn how to read a geologic map without any instructions in a class. It was my assignment to find those maps, copy them, and translate them. Mick drove and kept the car maintained. Russ and Kevin cooked. My brother just did his thing and did what we told him. Like a pack of long riders, we scoured the bleak deserts of California and western Arizona, hitting the legendary pockets supposedly still stocked with precious gems and minerals. We dug up baseball-sized geodes full of quartz crystals in the Potato Patch west of Yuma. Farther east we scrambled through the dank causeways of the Silver Queen mine and came up with some very decent specimens of sparkling copper minerals—green malachite and blue azurite. During a broiling afternoon we crawled down the shaft of a mine looking for giant yellow crystals of wolfunite. The crawlway had an overpowering smell of urine, and the source of the stench, a battalion of bats, spiraled up from the bowels of the mine and missed our ducking heads by a hairsbreadth. One of our most curious destinations was in the northern Mojave of California, a mine in a tawny gully called Last Chance Canyon. The name was inspired by a history of desperation and futility. Many decades before, a small clan of prospectors and their families had chopped a horizontal tunnel for about half a mile through a mountain and right out the other side, failing to encounter a single speck of gold. Allegedly the leader of the enterprise committed suicide, and the clan deserted the miserable site. The mine was now yielding a meager revenue; for a couple of dollars you could rent a carbide lantern, enter the mine, and retrace the path of this tragic folly of an excavation.

Our explorations culminated in what was nearly our own tragic folly. On a frigid, blustery late November evening we camped in the Bristol Mountains, a rugged no-man's-land about halfway between Los Angeles and Las Vegas. Our Thanksgiving dinner was a half-gallon can of Chung King chop suey. We had dragged that poor VW over cobblestone streambeds and plowed it through sand hills. When we at last found our deserted target of exploration, we were not welcomed. The mine was boarded up, its main entrance wrapped in barbed wire, and its subsidiary entrances either blocked with cement or even blown up. The requisite signs—DANGER, ABSOLUTELY NO TRESPASSING, TRESPASSERS WILL BE PROSECUTED—

were shot full of bullet holes. The next morning, with no lack of stupidity, we struck up our carbide lanterns and nervously entered this hellhole. Much to our dismay, the mine was a shaft that went several hundred feet straight down, with collateral, horizontal shafts intersecting this main pit every hundred feet or so. The only access to this underworld was via an ancient ladder with an incomplete set of rungs. Hours later and 300 feet down, our carbide lamps blew out in the damp drafts. In vain, I tried to light the flint of my lamp, sweeping it across the denim of my jeans. In the darkness, Russ, who was leading, yelled up to us.

"We got a bad rung, then three rungs missing."

To my astonishment, Mick, the most obsessive of us gold seekers, shouted back, "Can you make it over the gap?"

I heard heavy breathing and some creaking of the ladder. I could feel by the vibrations of the ladder that someone was moving below.

"I got my lamp back on. I'm going for it," Russ shouted. I couldn't believe it.

Then I heard a low groan, followed by a crack like a piece of juniper exploding in a campfire.

"Russ, are you OK?" several of us shouted in unison.

"Shit—this is bad. The rung's gone." Russ's voice had a shake to it I'd never heard before.

At that moment, the rung on which I was perched started to crack. I shifted my weight quickly and swung too far out at the edge of the ladder. My feet disappeared in the darkness. The fire from Mick's lamp, some thirty feet below, looked like a dying star. As I regained my balance and a toehold where the rung seemed more stable, my knife slipped out of my pocket. We heard it plunk into the mysterious pool at the immeasurable bottom of the shaft. I was trembling so much I could barely hold on. A long conversation ensued.

"Maybe we can make it," Mick said with fanatic resolution.

We each took our turn descending the ladder and inspecting the impossible gap that blocked our descent.

Finally, even Mick admitted it was hopeless. "Let's get out of here!"

The prospect of retreat was terrifying, however. Our carbide lamps were now either out or burning with a puny wisp of a flame.

"Mike, you're leading, do you want my lamp?" Mick yelled.

"I think I can feel my way up," I shouted back. Even in the pitch black of the mine shaft, I could feel how weak and rotten many of the rungs were. I ascended at a glacial pace, holding my breath with each lunge to the rung above me. I had read that most climbers die on a mountain during the descent. How odd, I thought, to die on this ignominious ascent. After an immeasurably long time crawling up the ladder, I noticed something strange. The rung just in front of me was crudely outlined in a pale green-white glow. As I moved up a few feet the light got better. I could actually see very faint flecks of grain in the rotten wood. Where was the light source? I looked up.

"There's a light up ahead!" I shouted.

"Where?" came the collective and excited question.

"I think . . . I think it's in the entrance of the shaft." I wasn't quite convinced myself.

What was that light? It looked like a car headlight or a high-powered flashlight shining down on us. Was it the light of a rescue party? Police? Bandits? At this point, any option was preferable to a 500-foot plunge down a forgotten mine shaft. I was getting closer, moving with more speed and confidence up the illuminated ladder. The perfect orb of the light had some peculiar wispy streaks on its lens.

"It's the moon!" I shouted.

There was no reply. My mates thought I had become entirely deranged.

But I was right. We had entered that mine at nine in the morning. We had dangled on that ladder in the darkness for eleven hours. On its providential arc, the full moon was now perfectly aligned above the entrance to our mine shaft. It was almost enough to make me go back to my catechism.

I crawled out of the hole in the blinding light of the moon, repressing tears of relief for surviving such a stupid enterprise. Later, I stood by the campfire shivering, and threw my sleeping bag around me like an Indian blanket. The moonlight had converted black cliffs to alabaster temples. I made some silent vows. No more mines, at least not ones with creaky ladders and vertical shafts. Epiphany is useful especially if it comes early in life. I had emerged from the bowels of the earth into a moonlit desert, the same one I crossed with my family on the way to the Grand Canyon, and the desert before me was the most beautiful, welcome place I had ever seen. Next time I would come back better prepared for exploration.

4

. . .

The Plateau of the Dragons

"Would you like to go in the field?"

The question came unexpectedly as I stood at the threshold of Dr. Peter Vaughn's office in the Department of Zoology at the University of California at Los Angeles. He was barely glancing my way, more intent in staring down at a Frisbee-sized piece of mauve-colored sandstone flecked with the white shimmer of fossil bone.

"Uh, yeah, sure," I stammered. I was beginning to feel the shock of this extraordinary offer. I was still a teenager, marking time in school to stay out of the draft, hardly staying awake in class after all-night binges of rock music. I played bass guitar in a band that took jobs anywhere we could get them—keggers hosted by brawling, thick-necked frat boys at USC, rhythm-and-blues dives in the edgy section of Watts, gay bars in Santa Monica, lesbian bars on Melrose Avenue, beach parties for rich kids, community recreation centers, even sock hops at our disdained alma mater, Saint Monica High School. I didn't have much time to hit the books. I certainly didn't warrant an invitation to join a professional paleontological expedition. Besides, it was 1969, the band needed to break through or break up; we'd all be twenty soon. We had some feeble plans for a summer tour that would get us out of L.A., and there were rumors that a big music festival in upstate New York would be worth a pilgrimage. But as I contemplated these commitments, my cortex flashed with the glint of red rock in the Grand Canyon.

"I'm taking a small crew this year to mop up. I plan to make a last stab at the Abo and Yeso formations in New Mexico, then up north a little way to the Caballo Mountains. We'll spend much of the summer in Monument Valley, but first there's a small quarry I'd like to open up near the Sangre de Cristo Mountains. I'd really like to find more of those amphibians." Vaughn spoke in a fast and challenging way, as if you were always his equal, as if you shared his prodigious knowledge of place-names and concepts. I later realized this was part of his brilliance as a teacher. He didn't expect you to know all those things; he wanted to show you what it would be like, how pleasurable it would be, to know them when eventually you got to know them yourself. He was now flicking the surface of the rock with a small needle projecting from a pin vise. He smoked a pipe, and he had a habit of reaching toward the corner of his mouth, even when he wasn't smoking, as if the pipe were still there.

"I feel a bit of a traitor to shift from the Permian to the Pennsylvanian, but those amphibians are really *neat*." He emphasized his old-fashioned slang with much enthusiasm. He was now time-tripping. The Permian Period of the Paleozoic Era ranges from 290 million to 250 million years ago. The Pennsylvanian is older still, between 320 million and 290 million years ago. Some Permian creatures were big, dragonlike, and charismatic, like the fin-backed *Dimetrodon*. Pennsylvanian-aged creatures were less imposing. They were early experiments in terrestrial lifestyles, probably like modern amphibians, frogs and salamanders, still tied to the water because their eggs lack shells that would keep the gelatinous ooze of yolk and surrounding membranes from drying out and killing the growing embryo. Many of these Pennsylvanian amphibians looked

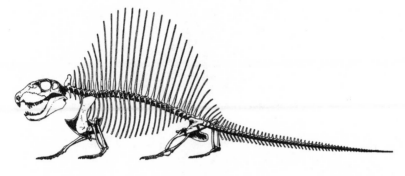

Dimetrodon

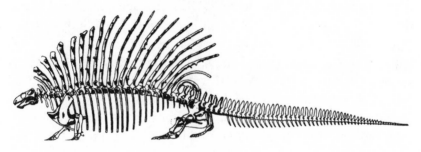

Edaphosaurus

like fat, swollen salamanders, sort of like Jabba the Hut. Others, like nectridians, had slender salamander bodies with boomerang-shaped heads. Their eye sockets—technically called orbits—converge toward the front of the skull; this made nectridians look cross-eyed.

"Of course, you have a driver's license." Vaughn squinted and picked up another mauve piece of Monument Valley rock.

"Yes," I lied.

Didn't everyone in L.A.? My procrastination about this was a mark of social retardation.

Vaughn fortunately believed me. "We'll take the green truck and a jeep from the University motor pool. They hate what I do to vehicles. It's rough country, especially those mesquite-filled canyons in southern New Mexico. Don't worry, you'll learn to drive with a granny gear."

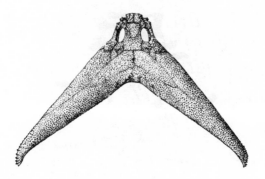

Diploceraspis, a nectridian

Vaughn had a phone call and asked me to step out for a moment. The courtyard outside the UCLA Life Sciences Building was flooded with amber late afternoon light, southern California light. Eucalyptus leaves fluttered like silver-winged butterflies in the soft west wind from the ocean. I thought of the sensual but ephemeral pleasures of summer in L.A. At that moment, I knew I was ready to forgo the seductive ennui of this So Cal ambience for a hot, dry, mesquite-filled canyon. But first I'd better get a driver's license.

When I returned to his office, Vaughn was bent intently over another small rock cradled in his hand. "Of course, I'd love to find some really good *Dimetrodon* this summer. I have only a few scraps so far." *Dimetrodon* was a dominant predator of the late Permian. It looked like a bloated lizard, about three and a half meters (eleven feet) long and weighing about 150 kilograms (300 pounds). It had a huge head and jaws with knife blades for teeth and a set of tall spines extending upward from its vertebrae. Although there is no direct fossil evidence, it never occurred to paleontologists that those spines supported anything other than a fanlike web of skin, something like the biblike skin flaps under the chin of some lizards or the webs between the fingers of tree frogs. Some lizards use a webbing as a wing or a gliding surface and float from tree to tree. *Dimetrodon* did not glide with its vertical fin. The fin was a single, flat, sail-like structure used for—what? We don't really know. It is popular to suggest the fin was something of a radiator, allowing these beasts to lose heat. Heat was a Permian thing by evidence of the geology, at least in North America. Permian rocks in places like

Utah and New Mexico are red beds thought to have turned red when their iron-bearing minerals were oxidized in the oven-dry air of an ancient desert. Vaughn had published some papers elaborating on this theory, even suggesting that some ancient mountains to the north and east deflected incoming rainstorms and put Permian desert territory in a rain shadow. This is much like the present-day situation in the parched Owens Valley in California, which lies just east of the Sierra Nevada. When the cooler coastal fronts drift eastward across California, they mix with warmer air and form big clouds over the Sierra. There they drop most of their water, depriving the low valleys to the east of any more than a few inches of rain a year. Thus *Dimetrodon*, like those fat chuckwalla lizards, had to contend with arid Owens Valley weather. So the fin was probably convenient for cooling off the monster. Tiny blood vessels formed a complex lattice under the surface of the skin, releasing the heat carried from the body's core. Or maybe the fin had some ritualistic function; bigger fins in bigger males may have attracted more females or may have scared away rival suitors. Both of these seem like decent speculations.

Dimetrodon is a dramatic animal, but it isn't a dinosaur. The Permian hosted many interesting beasts, but these died out at least 50 million years before dinosaurs appeared in the next time interval, the Triassic Period. As odd as it may seem, *Dimetrodon* and other less knarly "fin-backs," like the plant eater *Edaphosaurus*, are more closely related to us than to dinosaurs. They are an early offshoot of the big branch called synapsids, characterized by the presence of a large opening for the jaw musculature low near the back of the skull. Synapsids are a long chain of relatives that include species leading up to mammals. Accordingly, synapsids were traditionally and colloquially known as "mammal-like reptiles." But this is a bad name, because our best estimate of the relationships among land vertebrates suggests that mammals and their ancient synapsid relatives are found on an isolated branch outside the group that contains turtles, lizards, snakes, crocodiles, dinosaurs (including birds), pterosaurs, and the extinct marine plesiosaurs and ichthyosaurs. In the last few years it has become acceptable to call this latter group reptiles, and thus *Dimetrodon* and other extinct synapsids might

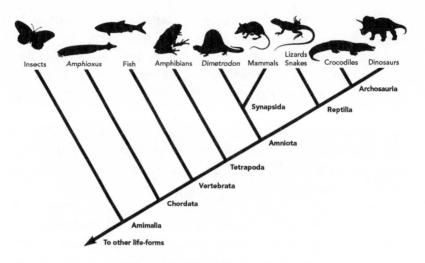

Insects   Amphioxus   Fish   Amphibians   Dimetrodon   Mammals   Lizards Snakes   Crocodiles   Dinosaurs

Archosauria

Synapsida   Reptilia

Amniota

Tetrapoda

Vertebrata

Chordata

Amimalia

To other life-forms

The major branches of the vertebrates

be better known as "reptile-like mammals." For these reasons, *Dimetrodon* and kin are important to some key questions about the anatomy and evolution of the vertebrates, and Vaughn studied them with passion.

"Have you ever been out there, Mike?"

"Only to the Grand Canyon, and only across the country, mainly at night, a few times with my parents . . ."

Vaughn looked away with a squint as if he were gazing at the north rim of the canyon on the horizon. "I like the Four Corners region. Fossils aren't always plentiful, but it's damn lovely. It will be hot as Hades down in southern New Mexico, but when we get up north into Utah, up in the center of the Colorado Plateau, it'll be cool. It'll feel like Alaska."

The Grand Canyon, where I first encountered a sizable chunk of the history of the earth, is only one cavity—albeit a very impressive one—in a great expanse of canyons, buttes, cliffs, and heaps and heaps of rainbow rock, an exquisitely carved terrain that stretches for half a million square kilometers (300,000 square miles) around the Four Corners (where nearly rectangular Utah, Colorado, New Mexico, and Arizona meet). Many of the earth's greatest natural wonders are located in this region, which extends from a corner of Nevada through southern Utah, the western third of Colorado, and northern Arizona and New Mexico. West of the Four Corners stands Monument Valley and its spires, which look like colossal chess pieces; farther west is the mighty Glen Canyon, now covered with man-made Lake Powell, sacrilegiously named after the explorer who so revered the once-unflooded maze. Nearby are great sandstone arches and, farther north, more tortuous canyons where the Green and Colorado rivers merge. To the east of the Four Corners are the magnificent chromatic cliffs of southern Colorado and northern New Mexico, where dinosaur-impregnated beds form a backdrop to the Spanish missions and their cemeteries with leaning crosses. The glories of exposed and eroded rock are bewildering and varied in their numbers. It's hard to contemplate a region that includes all the above and more—the monolithic sandstones of Zion, the sand castles of Bryce Canyon, small pockets of boulders and balancing rocks like Goblin Valley and the Valley of Fire, the great creases formed by cliffs at Water Pocket Fold, loops of huge rocks cut by the Green River at the Goose Necks, and the great rock citadel crowned with juniper and pine of the Aquarius Plateau. The rocks dominate, and their pervasive presence inspires a lengthy litany of often repetitive place-names. Every state has one or more of a Marble Canyon, Sand Canyon, Box Canyon, Rattlesnake Canyon, Pine Canyon, Stinking Water Canyon, and Red Rock Canyon. And for every Bone Canyon there is a Bone Wash, a Bone Ridge, a Bone Cave, a Bone Gully or (in New Mexico, southern Arizona, or northern Sonora) Huesos Arroyo.

All these place-names, whether redundant or unique, characterize a region that comprises a greater hodgepodge of landforms than anywhere else on earth. The complexity hardly seems to justify recognizing that the region has any unity at all. Yet geologists, with their eye for the big picture, see it: they call the whole mess the Colorado Plateau, after the mightiest of its meandering watercourses, which cuts a sinuous crevasse from northeast to southwest through the region. *Colorado* is the Spanish word for red, and one might claim that the predominant hue in this landscape is red. But the real diagnosis for the region that makes geological sense comes with the second part of the name—Plateau. Across most of its acreage, the Colorado Plateau is essentially a rather flat, elevated

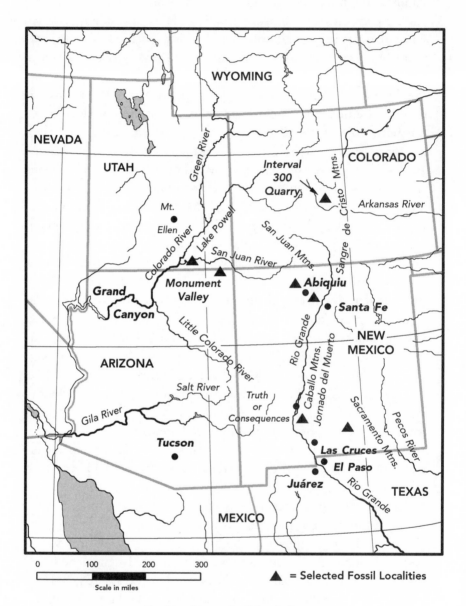

Fossil localities on the Colorado Plateau

plain. The canyons that plunge to abyssal depths and the mountains that poke up to the stratosphere are mere anomalies, products of powerful water and wind erosion and violent uplift which jostle the surface. And, like many plateaus, this one is impressively high, on average about 4,000 feet above sea level. When you drive along the plateau, you can't tell this because it is just too big. It's not like a lofty flat-topped rock tower sticking up from the Amazon jungle. But it is a plateau nonetheless.

Naming and diagnosing something does not mean you have explained it. Explanation requires a knowledge of earth processes. And this in turn requires a geological study of motion. The movements of the crustal plates and the upheavals they cause are phenomena formally known as plate tectonics. The word "plate" refers to a big section of the top layer, or crust, of the solid earth. There are roughly a score of these plates now mapped out for the globe, some huge, some smaller. The Eurasian Plate, for example, includes all of Europe and large parts of Asia northward to the Arctic Ocean. By contrast, the Cocos Plate is a parallelogram of crust just west of Central America that represents only a relatively small patch of sea bottom. The plates often coincide with more familiar features

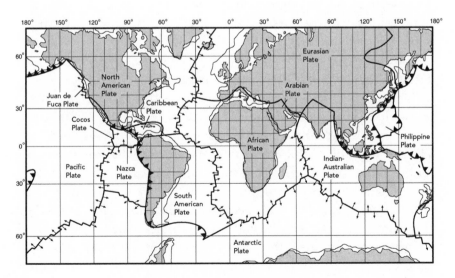

The major tectonic plates

of the earth's surface, such as the continents, but a single plate can include both a piece of continent and a piece of ocean floor. Unlike the continents, plates are not bounded by coastlines or even necessarily by the wide swath of shallow ocean fringing the land called the continental shelf. They are instead bounded by deep ocean trenches and by mountain ranges on both continents and ocean bottoms. The boundaries of these crustal empires are important because they are where the action is. This relationship between plate boundary and action warrants the second term, "tectonics"— geologists' parlance for the big movements, the quakes, shakes, buckles, and shifts of the earth's crust. Plate tectonics refers to those big movements at plate boundaries.

Earthquakes and volcanic eruptions are dramatic evidence of tectonics. But what geological forces do those violent events actually denote? We may consider earthquakes and volcanic eruptions as destructive—and they certainly can be when it comes to human life and property—but such upheavals are also constructive, for they are symptomatic of the forces that construct new pieces of crust and reshape the earth's surface. New crust is actually created through volcanic eruption and lava flow usually emanating from mountainous regions on the ocean floor. Crust spewing forth from these vents inches its way, like an excruciatingly slow conveyor belt, along either side of these spreading centers. One can visualize something like a crack in a house floor where water seeps in and spreads symmetrically on either side of the crack. It may be counterintuitive, but "solid" earth crust sometimes becomes fluid.

One may naturally wonder what happens to the new crust as it spreads out on the ocean floor. After all, the earth has a finite circumference and diameter; the spreading of crust at the sea floor has not added to the size of the earth over eons. So what happened to all the extra crust? (There are some controversial theories that the earth has actually expanded through time because of the accretion of new crust, but almost all geologists reject this claim as overly speculative and inconsistent with the available evidence.) The key to answering the question comes from one of the most stunning observations in the history of geology, made in the 1960s at the same time that I was just learning to recognize rock types in a beginning geology lab. After careful analysis of the age of many sediments, it was observed that nowhere is the ocean bottom more than roughly 200 million years old. Now obviously, much of the earth's crust on continents contains much older rocks, such as the 1.7-billion-year-old rocks at the bottom of the Grand Canyon or those 3.5-billion-year-old ones in Greenland. There had to be an explanation for the youth of the ocean basins and that explanation was beautifully simple: the older oceanic crust had simply disappeared, it had somehow been "eaten up" somewhere. It was subsequently found that the deep oceanic trenches that abut volcanic islands and the mountainous shores of continents are the places—the plate boundaries—where crust is consumed.

The juxtaposition of trenches with volcanic islands or mountain ranges at the edge of continents is not a coincidence. As the oceanic crust is consumed and moves deeper into the trench, it is squeezed and heated by the pressure of thick continental crust above. The deep, hot, molten rock bulges out, pushing and distending the overlying rock. In places, it escapes through cracks to the surface. If it stays hot, it emerges as lava or explodes in furious clouds of sulfur, carbon dioxide, and ash from the vents of volcanoes. If it cools before it reaches the surface, it forms huge tombstones of smooth granite and other rocks. Rock masses of this kind—cooled bodies of once-molten rock—can be seen in the places where softer rocks above them have been eroded away. Much of the Sierra Nevada—and Yosemite—is made up of such masses, called batholiths ("deep rock"). Some of the greatest unbroken precipices on earth, like Yosemite's El Capitan, are batholiths. Thus, the eastward plunging of the crust on the bottom of the ocean, pushing below the leading edge of the continental crust drifting westward, explains the origins of California's coastal ranges and Sierra Nevada. Additional symptoms of all this squeezing and reworking are earthquakes, the uncooperative groans that emanate from deep within the earth when stable crust is forced to become more fluid.

Plate boundaries that primarily involve trenches are called subduction zones, because these are sites where crust is being carried down, or subducted. Forces at these boundaries are due to compression, as big masses of molten rock push together with great momentum. The earthquakes that result from this compression usually spread from points very deep within the crust; they are called deep-focus earthquakes. There are marvelous tectonic maps which show that deep-focus earthquakes occur largely in or near the great trench systems of the world and the great mountain chains next to them.

The other major plate movement involves spreading as new crust is generated or as plates move apart. This spreading is caused by tensional forces, for example when crust is generated at oceanic ridges, or mountain chains on the sea floor. Plates move apart in many other places; often the movement takes the form of two blocks of crust sliding past each other. The tensional forces behind this sliding create cracks of weakness in the crust—faults, like the 1,500-mile-long San Andreas Fault between western and eastern California. Here, earthquakes tend to occur at less depth than the earthquakes commonly associated with trenches, and they are called shallow-focus earthquakes.

One can easily see why the prediction of earthquakes is not so haphazard or mystical. Earthquakes are more likely to occur where the geological action is—where there is sufficient sliding, spreading, faulting, subduction, and eruption of big masses of crust. But plate tectonics can tell us more than this. It can explain the birth of many mountain ranges, like the Sierra Nevada, the Alps, Andes, and Himalayas, and the deep trenches in the oceans that border places like the Pacific islands of Tonga and the Marianas and the western edge of South America. Plate tectonics tells us that the mightiest mountain range on earth—one nearly completely submerged underwater—the 40,000-mile-long Mid-Atlantic Ridge, is the source of new flooring for its namesake ocean. But plate tectonics is insufficient to explain all features of the earth, including the Colorado Plateau. A conveyor belt of plunging oceanic floor would have had to thrust downward at too shallow an angle to account for a piece of this elevated crust that is 500 miles wide from west to east. It would not have created enough pressure to heat the rocks and raise the crust. Besides, the Colorado Plateau lies too far from the western margin of North America, which marks the boundary between the North American Plate and the Pacific Plate. It is far from the action.

Finally, it is a mystery why this vast region rose rather uniformly to such an elevation without much trauma. The crust of the Colorado Plateau is largely devoid of the overturned rock and violent upthrusts of slabs that typify mountain ranges like the Andes. Such a major piece of deformed land is the exception rather than the rule. In contrast, the earth's highest plateau, the one that encompasses Tibet and much of western China, was formed with the uplift of the Himalayas on its southern border. This raising of the "roof of the world" occurred when a northward-drifting India collided with what was then the southern edge of Asia—the annealing of subcontinent with continent—some 40 million years ago. But no such gargantuan collision can account for the Colorado Plateau. Geologists recognize a number of factors that could have all contributed to its formation. These include the collision of the Pacific Plate with the North American Plate to the west, the rise of the Rockies to the east, and the extensive volcanic activity that accounts for the great steam basins of Yellowstone to the north. Yet these different explanations cannot be well coordinated to form what is essential to the progress of science—a simple answer. The Colorado Plateau remains a lofty, inert, and exquisitely sculpted land of mystery.

Fortunately, you don't need to understand earth processes intimately to reap their benefits. The regular accumulation of layer after layer of sediment on the Colorado Plateau has maximized the chances for the preservation of fossils there. Fossils need quiet beds in which to take a rest. Overturning, faulting, compressing, and reheating rocks will deform or shatter whatever fossils they contain. That is why the most famous sites for gorgeous dinosaur skeletons, ancient bird fossils with feather impressions, and fish that are still perfectly coated with scales occur in ancient lake beds, quiet meandering rivers, mudflats, or stabilized sand dunes. Many of these quieter environments for entombment characterize the rocks of the Colorado Plateau.

But this quiescence in sedimentation is not enough to provide good fossils. There are plenty of layers of quiet rock which nicely preserve fossils that will never be seen, like those deeper layers of shell beds in the Santa Monica Mountains. They are covered not only by vegetation but by blankets of overlying rock. Something has to bring all these riches to the surface. It's true that fossils have been found deep within caves, mines, and drill cores. In fact, most of what we know about microscopic fossils that once lived in the oceans comes from thousands of sediment cores that have been drilled and hauled up on the decks of oceanographic vessels. But dinosaurs, mammals, and many other fossils are almost never extracted this way. Instead they emerge from the rock through the cooperation of wind and water, the same forces of erosion that molded the Grand Canyon and the rest of the Colorado Plateau. The stately rise of the Colorado Plateau and the slow but powerfully corrosive action of wind and water on its rocky surface make for perfect conditions of exposure. Contrary to some misguided portrayals in fiction and movies, paleontologists don't walk over the ground with a divining rod and suddenly point to a spot and exclaim, Let's dig here! The chances of finding exposed bone are of course increased with the amount of surface to explore in. And surface area is not just a function of the horizontal acreage of exposed rock. A cliff 500 feet high has more potential for exposing fossils than a cliff 50 feet high, and a 500-foot cliff carved into deep gullies, towers, and promontories has more potential for exposing fossils than an equally high cliff that forms an unbroken rampart. When prospecting for fossil bones you can walk up the gullies and cover a distance of thirty miles in a day and yet only be five miles from camp as the crow flies. That's a lot of surface area to cover. And it's hard to imagine a place that has more surface area of fossil terrain than the Colorado Plateau. The place has the right combination—layers of rock laid down under stately conditions, lots of rocky acreage, and lots of erosion.

One would think this makes a strategy for field paleontology a no-brainer. Why waste your time elsewhere in crummy outcrops covered with vegetation when you can explore these glorious canyonlands of the American West? But fossil exploration is not so straightforward. Two factors complicate matters. For one, in plenty of regions of the world where searching is imperative because of the potential importance of the sites, circumstances make them poorly suited for fossil hunting. Antarctica is virtually covered with ice several miles thick, but discoveries of dinosaurs and other fossils on the little blobs of rock exposed on isolated islands that flank this frozen continent are resoundingly important. These fossils offer facts about life at the southern end of the world millions of years ago, when Antarctica was a friendly place, lush with vegetation and broadly connected to the landmasses that eventually became Africa, South America, and Australia. There are other regions of great biogeographic importance—where fossil information would tell us vital things about the distribution, evolution, and even extinction of animals and plants—that are equally bad places to look for fossils. Many paleontologists spend an inordinate amount of time and money looking at such unlikely terrain as the rugged Greenland coast, jungles of the tropics, and terraced agricultural slopes in the Middle East, Laos, and Indonesia.

A second reason why paleontologists don't just target places with plenty of exposed rock is a simple sobering statistic about the chances for fossil preservation. Good rocks don't guarantee good fossils; in fact they don't guarantee fossils at all. Vaughn knew this problem well. Although he had found many fossils in various rock layers, he had also spent ten years scouring some of the most auspicious shale and sandstone deposits on the Colorado Plateau with nary a scrap of bone to show for this effort. Sometimes these experiences can be heartbreaking. A nearly complete skeleton of an important primitive amphibian, *Tseajaia campi*, was found lying on the surface of a slope below a cliff in Monument Valley. This brilliant specimen, the subject of a dissertation by one of Vaughn's doctoral students, was presumed to have weathered out of a rock layer, the Organ Rock Shale, that formed the lowest layer of the cliff just above the slope. Yet Vaughn and others prospected that layer intensively for years in vain. Where the hell were those other great fossils?

Vaughn seemed to recognize the importance of instilling realism at an early stage. He described paleontological fieldwork as 49 percent anticipation, 49 percent recollection, and 2 percent success.

At the time of his invitation to the field, the odds against paleontological success were hardly my concern. I had other challenges to deal with. In addition to driving with competence, I would have to demonstrate some practical skills—rope tying, truck packing, tent pitching, fire starting, route finding, rock hammering, excavating, fossil hauling, crate making, and even dynamiting. Suddenly the prosaic tasks of a bone hunter seemed harder to me than the most sophisticated geophysical equations that plate tectonic theory required.

5

. . .

The Rookies

In addition to the basic skills needed for my first real field expedition, there was another, less tangible necessity for survival. This was simply the ability to balance experience and common sense so as to avert disaster or humiliation. There was a kind of hazing then traditional with field rookies, a teasing and duping often aggravated by the victim's own naïveté or inability to adapt. I heard plenty of tales of such hazing from past field seasons.

One poor fellow, a premed major who decided to join Vaughn's crew to round out his education, had virtually no experience of life outside the city of Los Angeles. Early in the summer, he saw frequent signs on ranch fences—WARNING: CABLE CROSSING—that indicated the pathways of buried phone or power lines. When he asked a paleo grad student to explain the meaning of this admonition, he was told that "cables" were deranged, renegade bulls that broke through fences and raped defenseless cattle on neighboring properties. Cables were mean, violent, and not at all hesitant to gore humans. The poor fellow had much anxiety prospecting in areas where such signs were seen. He began to convince himself he could recognize a cable from a football field away. One day he jumped out of the jeep to open a gate between two cattle empires. He saw a putative cable sauntering toward him and ran for the hinterlands, leaving the gate wide open. The cable, which was in reality an old cow, lazily walked through the gate and invaded the forbidden territory. An angry cowboy came galloping down the hill, lassoed the errant cow, and nearly horsewhipped the fleeing gatekeeper. When Vaughn rendezvoused with his crew later in the summer, the first thing the tyro said to him was, "Sure are a lot of cables out here, Dr. Vaughn." The professor shot an annoyed glance at his grad students.

Other dangers were warned against besides "cables." "Cot-level bats" were said to swoop down and claw your face on 85-degree nights in southern New Mexico, unless of course you slept in terrorist regalia, with a suffocating red wool ski cap pulled over your face. There were also those excruciating beasts called *Psidocurum causatum*, commonly known as Schaeffer's beetle, which crawled up one's leg and inflicted a painless but insidiously disastrous bite in the genital area. Six months after returning from the field one supposedly experienced the worst nightmare of any reproductively viable male—the testes would atrophy and drop off. The Schaeffer's beetle myth was so elaborately cultivated that very authentic-looking notices warning visitors of the spread of this species were surreptitiously tacked to bulletin boards at ranger stations. At one station the perpetrators even got a ranger to play along with the joke. Allegedly the only protection at night from these nasty insects was to keep water between you and the attacker. In a flash of ingenious altruism, the hapless victim of this prank went into town and bought pie tins for everyone, for these could be filled with water and put under the foot of each cot leg.

I never witnessed any of these deceptions firsthand, and I didn't even know if the field tales themselves had any reality. But the legends were enough to instill apprehension. I had about four months to become more field-ready. I finally attained the biggest prerequisite, the driver's license, but this didn't mean I could maneuver a pickup truck up a steep hill using a granny gear. And what if I fell victim to pranks? What new fantastic danger would I succumb to? What cables, cot-level bats, and Schaeffer's beetles awaited me?

With some trepidation I met with Vaughn and the rest of the crew at his house in June 1969 to pack the truck that, along with our jeep, would form our caravan. Vaughn did this assiduously, with precise planning: each crate, cot, rope, tarp, burlap bag, pickax, rock hammer, chisel, watercooler, camp stove, duffel bag, and plaster bag had its rightful and unalterable place. Then Vaughn tied the truck bed down, threading the ropes through the grommets of the tarpaulin thrown over the entire lot with a series of impressive half hitches, bowlines, and slipknots. He did this with speed and grace, like a cowboy tethering a struggling calf. Not bad for a professor of zoology at UCLA, trained in paleontology and evolutionary biology at Harvard University. I wondered if I would ever be able to secure a tarp with such mastery.

We headed off the next day through the broiling, locust-filled Imperial Valley at the southern edge of the Anza Borrego Desert, aiming for a first night's rest in Tucson. At one point I was very casually offered the wheel of the stake-bed truck. At first I was a little shaky, but the straight flat ennui of the highway lulled me into confidence. A lumbering fuel tanker in front of us beckoned me to pass him. As I hit the gas pedal and veered into the oncoming lane I could see a semitruck coming at me from around the bend some distance away. How far away? Far enough, I thought. I gunned our truck but its speed remained constant. The oncoming truck now blocked the eastern sun, and its horn practically blasted us out of our seats. I cleared its cowcatcher by inches and pulled over to the side of the road, kicking up a tornado of gravel and dust. Vaughn's sixteen-year-old son, David, who had been condemned to ride shotgun with me, was succinct in his appraisal. "You're not a very good driver," he said with some agitation.

That wasn't the last trial of the day. As we drove through the endless scrubland of California's southernmost deserts the sun bar-

becued my left arm, which stuck out of the driver's window. When
we stopped to cool the truck down, you could feel the heat rising
from the boilerplate of the asphalt highway.

"Pretty hot," I said.

"Yeah, about as hot as southern New Mexico, where we're going," David replied.

In the late afternoon the heat did not really dissipate, but the
shadows cast by low hills of volcanic rock gave an impression of
some relief. We spied a field of great sand dunes that looked like
snow hills. Out of the trucks, we all took off, running up the sand
slopes like wild Bedouin, then whooping and cartwheeling down
the dune's steep leeward slope. The rest of the crew was satiated after one or two runs and took a break in the puny eastern shade of
the vehicles. My liberation from the driver's seat and my exhilara-
tion spurred me on. After the tenth roll in that hot soft sand I felt a
little faint and dizzy. Then I walked over to the back of the truck
and threw up.

"Watch the heat, Mike. You're dehydrated. Eat something salty
and drink something. Then relax." It was Vaughn's first lesson in
desert survival.

We left the sand dunes and plunged eastward into the suffocating
depression surrounding Yuma, Arizona. Here the Colorado River
creates its own tropics, as the dribble heading to the Gulf of California adds enough moisture to the 110-degree air to bleed out
your remaining body water into sweat pools on your skin and shirt.
Yuma is a bad place for trucks, too, and we stopped to cool ours on
the long upgrade of Arizona terrain east of the river. I looked back
toward California and could see the faint blue hills where Mick,
Russ, my brother Steven, and I had scooped up geodes in the gray
sands of the Potato Patch. Then I looked around at my current field
mates. Except for Dr. Vaughn himself, they didn't seem very different from that earlier gang; they were all young—Vaughn's son was
only sixteen. There was another young'un, Larry, son of a UCLA
ichthyologist, Dr. Boyd Walker. My contemporary was another un-
dergrad and close friend, Jon Kastendiek; he was a somewhat reserved and contemplative fellow, who counterbalanced my more
outgoing style and my bursts of enthusiastic but sometimes rather

random activity. We were all rookies. Possibly we would be joined later in the summer by some of Vaughn's seasoned grad students. Then the hazing would likely begin.

Vaughn, at forty—our elder, our mentor, and our guide—had a low-key way of introducing us to this new world. His style in the field was in sharp contrast to his rather formal, quick, and challenging posture on the stage of a university auditorium filled with comparative anatomy students. It was as if the heat of the Arizona desert slowed him down to the metabolism of an inert iguana basking in the sun.

"Basin and range," he said simply, waving his arm out to the desert, to the deep valleys and the sharp-edged mountains thrust up by faults on the valleys' flanks. Then he got back in the jeep. Much of California, Arizona, and Nevada is a monochromatic landscape of white gypsum-coated dry lakes below black mountains dancing in the distant heat. The black ranges that confine the white basins are aligned north-south for hundreds of miles.

As we drove on, we rounded a curve on the shoulder of a mountain and looked down into a field of rocky cones.

"Volcanoes," Vaughn said. Then, after a long pause, "Young, boring volcanoes, only forty million years old or so." The disdain made sense. Vaughn's passion was 280-million-year-old rocks. He was fond of saying, "Everything important in vertebrate evolution happened before one hundred and fifty million years ago." You could see his point. Despite the drama of unfolding history during that last "measly" phase, virtually all the major groups of vertebrates— the sharks and rays, the ray-finned fishes, the lobe-finned fishes that branched off to land vertebrates, the amphibians, and the various reptile groups (flying pterosaurs, turtles, lizards, swimming plesiosaurs, dinosaurs, and crocodiles)—had all appeared by 150 million years before the present time. Even the tiny shrewlike fossils of our own group, the mammals, were known from Triassic rocks in England, South Africa, and China dated around 210 million years.

In more expository moments, Vaughn had a certain flair for sharing these lessons. He liked to quote arcane statements by Greek scholars, Roman generals, or his Harvard professor, the legendary Alfred Sherwood Romer. But he most often quoted, with an odd mix of admiration and sarcasm, John Wayne's lines in his best-known westerns. As we gazed upon the forbidding, uninhabited, infernal volcano field below, Vaughn applied one of Mr. Wayne's famous lines, "Sooner or later, boys, this land will be all fenced in," as if this rueful observation forecast an inevitable future human invasion. Of course, both Vaughn and John Wayne eventually turned out to be right. Two decades later, tract homes and golf courses that sucked up the feeble runoff of the Colorado River anastomosed like moon colonies all over the southern deserts of Arizona (my brother Steven even has a house there among the cactus). At the time, though, the prediction seemed absurd.

We spent our first night in Tucson in an undistinguished motel under the eucalyptus trees along Speedway Boulevard, an ugly strip lined with gas stations, used-car lots, and auto-part stores that seemed a perversely authentic imitation of the similar, depressing boulevards in L.A. But the Mexican food was good, and a bit more spicy hot than the L.A. variety. The morning brought another long drive through the basin and range, the saguaro cactus, and the creosote bush of southern Arizona. By early afternoon we reached Las Cruces, New Mexico. The town's white adobe houses looked like freak snowflakes spread out below the lofty, coxcomb summits of the Organ Mountains. From there we drove north and east into desert country sliced by steep-walled drainages trending westward from the Sacramento Mountains that cut through the Permian Abo Formation.

When we stepped out of our vehicles into the hot blast of the afternoon, we had a vivid validation of David Vaughn's prediction about the temperature. The canyon seemed to be drawing in all of the solar heat of the planet and bouncing it off its red Permian rocks. We made camp on a rim of sandstone sprinkled with mesquite and creosote above the oven of the canyon. Then we set upon the rocks for bone.

6

. . .

Journey of Death

True to form, Vaughn didn't give much of a briefing for our first day of field prospecting. "Just walk around and look," he said. "The bone is blue-white, just like back in the lab." Like everyone else in the small crew of five, I struck out alone. I went along, looking at the ground, moving from one gully to the next, avoiding clumps of mesquite and prickly pear cactus. It was hot—that's all I could really think about, except that maybe this job wasn't for me. This went on for several days. I don't remember much else of the experience. On the second day I scrambled in a wide circle over the purple-red Permian rock I had traversed the day before, hoping to chance upon a fleck of bone I might have missed. Vaughn often repeated a field law: never walk over a piece of ground only once, even only twice or three times; you haven't walked it until you can name every pebble on its surface. True enough, the second time I walked this plot I was dazzled by the sight of a blue-white fragment of a pelycosaur spine—my first substantial discovery—only to find a chalky, circular scratch from Vaughn's hammer already incised around it.

The name Permian, for the age of the rocks exposed on the walls of the canyons, comes from the name of a town in Russia. The major periods of the geologic calendar are named after the places where rocks of that age were discovered early in the science of stratigraphy and are particularly well exposed. The rocks exposed outside the village of Perm were thus the reference sequence for rocks, or type sections, for the Permian Period. I knew this history when I walked the Abo and the Yeso, and I imagined that the Permian of Perm was much cooler and more inviting than the Permian of New Mexico.

On the third day I staggered up a side of the canyon where the rock walls were higher and deeper and the boulders more plentiful. Out of sight of the rest of the prospectors, I could duck into the shade of a big boulder for momentary relief. As I did this, with some stealth, I heard a distant baritone shout.

"Hey, Mike—come here." I was mortified to see Vaughn standing about fifty yards away on the rim of the canyon. What was it? A complete pelycosaur skeleton that I might have overlooked in the path of my retreat to the shade? I expected a lecture on the need for field perseverance, but instead Vaughn merely smiled and pointed to a distant bend in the canyon that was highlighted by a splotch of chartreuse. "Come on," he grunted and sprang down-slope. Soon we could hear a distant sibilance like that of a garden sprinkler. We came around the shaded bend and saw a miracle of Lourdes—a spring spilling out of the rock in a radiant waterfall over a light green tapestry of maidenhair fern. The water pooled under the waterfall, formed a reflective ribbon in the wide, dry bottom of the wash, and disappeared into the gravel a few hundred yards from where it had miraculously emerged. Vaughn was laughing, filling his shirt, and dousing his head with the astoundingly cold water. I drank the spring water until it distended my stomach. It was delicious, despite a slightly gritty taste and a smell like wet plaster. Vaughn, I noticed, drank less.

The maidenhair fern spring, naturally enough, became the favorite stop for the entire crew during our remaining days in the canyon. But a few days after its discovery, we all noticed that our digestive tracts had become highly volatile. Vaughn noted that this was to be expected from all the slurping we did at our waterfall.

"That water flows over minerals that work just like milk of magnesia on the gut."

On some days we took driving expeditions. We liked the break from the routine of the long lonely walks, but I was terrified that I would be called upon to pilot the jeep on one of these forays. This would be my first real test in off-road skills, and the terrain was not a good training course. The boulder-strewn bottom of the canyon was punctuated with huge stands of mesquite. The thorns of this bush damaged our jeep tires. The day of reckoning finally came. As I ground through the gears of the jeep, Vaughn was kind enough not to remark on my driving. The others were less charitable and I hardly blamed them; they realized that in his generous mentorship, Vaughn was putting the jeep, himself, and his crew in danger. As we entered the narrows of the canyon, Vaughn ordered me to stop the jeep. In a flash of courage, he jumped up on the hood and seated himself with his legs dangling over the front grill, like John Wayne on his Safari truck in *Hatari!* As I inched cautiously forward, Vaughn peered ahead like a spotter on the bow, pointing left and right to threatening rocks ahead. Soon I started to relax, and I eased up on the brake, downshifting between second and first to take a big rut or a boulder. "That's it," Vaughn encouraged. At the very moment when I began to feel at one with the canyon, I plowed into a monstrous mesquite. Then I saw a man with a ripped shirt and a bleeding chest on the hood of the car. I stopped the jeep and ran around to the front in great agitation. "Are you all right, Dr. Vaughn, are you all right? Oh God, I'm sorry!"

Vaughn was groaning, but despite the blood and his pursed lips and furrowed forehead, his eyes betrayed a smile and a chuckle. At last he had the perfect opportunity to quote one of his favorite lines from *Red River*. In a gravelly, pain-inflected voice he said, "Never apologize, son. It's a sign of weakness."

Soon, after more than a week and only a few bone scraps to

show for it, even Vaughn had had enough of the frustrating fossil poverty. "Let's go to El Paso," he announced unceremoniously.

"This is sort of a break, and sort of a tequila run," David later confided, noting that El Paso gave easy access to the magical liquid derivative of the agave cactus which was sold at huge discount across the border. What little I tasted of tequila had not impressed me—it smelled like oil, went down the mouth as a film, and burned the throat—but a change of scenery had an overwhelming allure. We eagerly pulled out the stakes and guy ropes around the tents and packed the truck with great efficiency. Then we crested the utterly cool pine forests atop the Sacramento Mountains, the gateway to the long but fast drive to the Big Bend country of Texas. On one of those fair summits we scaled a 200-foot vertical ladder on a ranger's fire tower. This ascent somewhat unnerved my mates, but I had no trouble scampering up a ladder with pristine, dependable rungs fully defined by broad daylight. From that perch we could look west and see the pale pinks and browns of the Abo and Yeso formations—the place of our epic struggles with heat, dust, and barren rock. That terrain of torture looked puny and un-challenging from our elevation and distance. Vaughn gave a typical clipped lecture on the geology of southern New Mexico. Then we hit the road.

We spent our days in El Paso making forays into Juárez to pick up one-liter bottles of tequila and walk them back over the bridge that spanned an emaciated Rio Grande. Then we headed back to-ward Las Cruces and shifted north to the town of Truth or Conse-quences, or, as the locals called it, T or C. Vaughn commanded that we stop here to fix some of the mesquite-riddled tires. Lulled by a couple of days of icy drinks and motel swimming pools in El Paso, we gladly complied—anything to delay the next struggle in the broiling Permian. We cruised this settlement of Tastee-Freezes, A&W Root Beer stands, and vintage junk stores, rummaging through the latter for old cigarette lighters, playing cards, and an-tique comic books. Our favorite prizes were bits of glass and glass bottles, everything from abandoned opium flasks left by Chinese railroad workers to the coveted insulators from vintage telephone poles that age purple in the sun. While Vaughn kept an eye on the gas-station jockeys, making sure they plugged all the punctures, we hung around the town square, where sunburnt girls in halter tops and tight jean shorts came up to us. "You guys new in town?" they queried with a giggle. "You're from California?—thought so." "Laake thet haar," one of the girls said, and extended her arm to tousle my mop. We tried to be quiet and cool, but we liked the attention.

The T or C town square was not a pastoral paradise; it was adorned with a few ragged poplars and stunted tamarisks, but it also offered the most important monument to local history: among the dust-caked geraniums was a small bust of Colonel John Mosby, the notorious raider who chased Union troops through this valley in skirmishes of ultimately little consequence to the outcome of the Civil War. The bust looked oddly familiar. As Vaughn noted, the effigy clearly showed that Colonel Mosby could pass for the identical twin of Kirk Douglas, with the serious countenance of the firm-jawed slave Douglas played in the movie *Spartacus*. Hoping for more explanation, I mentioned this observation to the old lady who owned the glass bottle store. "Who's Kirk Douglas?" she asked, with a look of astonishment.

By the late afternoon, we heaved the patched tires into the truck and pointed our wheels toward the Caballo Mountains, which rise east of T or C and the Rio Grande as a parched, brown rampart that looks like a crumbling adobe wall. *Caballo* is the Spanish word for horse, but there was nothing reminiscent of a horse in the profile of those mountains. Perhaps Comanches once corralled their warhorses in the upper reaches of the range's canyons. Vaughn didn't know the reason for the name, and, maybe because he knew so much, he always freely admitted ignorance. Crossing the range, we descended its eastern escarpment and saw a broiling flat of white gypsum and sand with another much more evocative Spanish name, Jornada del Muerto—Journey of Death.

We occupied a forlorn cluster of deserted ranch houses and corrals on the edge of that hellish valley. Our only co-inhabitant was a quiet Mexican cowboy, Rodríguez, a fence rider who spoke no English. He lived in a small room in one of the decrepit buildings of the ranch. When I went over there to offer him a swig of tequila, he was slouched in a disemboweled old armchair, sound asleep, with a paperback novel on his lap that seemed the Mexican equivalent of *Riders of the Purple Sage*. Like the antiheroes of those western sagas, Rodríguez had a whiff of mystery, like someone atoning for a tragic mistake, a gunfight or an armed robbery, and living out his life of contrition on the Jornada del Muerto. Indeed he seemed to ride out his penance daily over hundreds of miles. While we scrambled up Permian rocks of the Caballo Mountains we could look out over the burning valley and see a lone white dust devil kicked up by his mount on the salt flats. A few hours later, I would ascend a steep promontory and see him above me on a ridge, standing next to his horse, looking my way as if he were making sure I was all right. He'd nod his head, tipping his sweat-stained ten-gallon hat toward me, and ride off. Often in the late afternoon we would hear a shot bounce off a canyon wall two miles away, and we knew it was Rodríguez.

There were never more than two or three shots. And there were never less than two or three jackrabbits for dinner. Rodríguez hauled these back to the ranch, told us of our rabbit meal with the simple greeting, *"Conejo,"* skinned the carcasses in less than ten minutes, and unceremoniously handed them over to us. The first night of this ritual, we sank our teeth through the tough meat and expressed our faux delight. Because we overdid the gratitude a bit we were driven to try dozens of recipes for *conejo* on subsequent nights. We pounded the hell out of it, marinated it, boiled it, or slow-roasted it on a spit. After one or two nights Rodríguez politely refused our preparation and stuck with his refried beans. Even he could not tolerate *conejo* the way the gringos prepared it. It occurred to us that this might be some monstrous practical joke on the cowboy's part—perhaps he had been primed by some of Vaughn's evil grad students. The favored hypothesis, however, was that our cowboy had a secret recipe that rivaled the *lièvre à la royale* at the finest Paris restaurant, but he was too polite to impose his culinary advice on us.

Our fourth day in the Caballo Mountains was a particularly bad one. As on other days, we found only little flecks of *Dimetrodon* bones, but there were more flat tires and more rattlesnakes. Indeed,

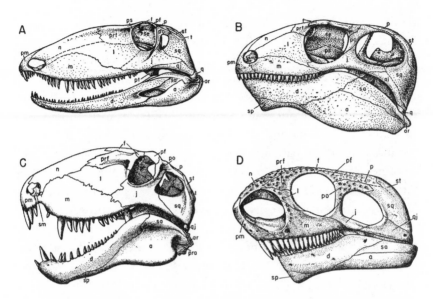

Skulls of pelycosaurs: (A) *Ophiacodon*; (B) *Edaphosaurus*; (C) *Dimetrodon*; and (D) *Cotylorhynchus*

I had to jump through a cave opening blocked by a couple of big diamondbacks. Vaughn sat down on a rock cairn where a freshly molted member of the same species lay coiled near the warmth of his rear end. On the way back to camp, one resolution lifted our spirits—"No *conejo!*" When I saw Rodríguez standing by the old barn we used as a base camp, holding a tether of dead jackrabbits, I almost ran him over. I barbecued the rabbit to black coal and made an extra-large side of macaroni and cheese. We pretended to eat the meat, but when Rodríguez slipped away to his room for his usual early sleep, we threw the charred carcasses in the back of the stake-bed truck. We knew we couldn't bury the rabbits because Rodríguez's dogs would surely dig them up. Instead, Vaughn and I drove resolutely for an hour to the far ends of the earth, knowing well that the cowboy was likely to cover half of New Mexico on his daily range ride. We stopped the truck near a small gully and jumped out to dispose of the rabbits, but were distraught to find them missing. There was little need for subtle deduction— Rodríguez's dogs had probably dragged them off the truck before we were even out of the gate of the ranch. Our cowboy host got the message, and that was the last of the rabbit meals. We regretted our impropriety but delighted in the new meal plan.

Perhaps I should write more of the Caballo Mountains, about their geology, fossils, and importance to science. But in truth, some ill-tasting jackrabbits, the heat, the mesquite-assaulted tires, the rattlesnakes, and a curious town with cute, sunburnt girls and a statue of Colonel "Kirk Douglas" Mosby are all I remember about the damn place. Doing paleontology, I was learning, can involve tedious stretches of time that seem as interminable as the Precambrian Era. After several weeks in southern New Mexico I began to wonder if Vaughn came out here only to quote John Wayne and eat *conejo*.

He may have begun to wonder, too, for he soon took us north to Colorado and Utah where, he claimed, there would be more fossils and it would be "as cold as Alaska." En route we stopped in Santa Fe, Vaughn's favorite town in the world. In those days as now, Santa Fe had its charming plaza and its stately Mexican adobe edifices with massive timber porches over plank walkways. But the place did not yet have pink-and-blue Southwestern *nouvelle cuisine* restaurants and parking lots full of white Mercedes convertibles. I agreed it was the most beautiful town I had ever seen. On a moonlit night we roamed the main plaza, where there was a big crowd and much fanfare over lunar rites. Television sets were rolled out of barbershops and curio stores and faced toward the plaza. In the flicker of a distant green TV screen, we saw Neil Armstrong get out of a lunar exploration module, walk down a silver ladder, and step into some soft dust that reminded us of the ashen rind on the flats of the Jornada del Muerto. We couldn't hear anything of his "giant step" speech except some static lost in the laughter and cheers of people in the square. We then followed Vaughn to a bar.

I was a curiosity in these establishments, not because of my youth but because of my long hair. Vaughn used to say, "You better walk a few yards ahead of us," when we roamed a sinister neighborhood in Juárez. Vaughn's favorite bar in Santa Fe was a dingy dive where a Mexican trio bashed out semi-rock versions of mariachi music, interspersed with a few awful renditions of Buddy Holly songs. The most pyrotechnic stage effect was a blinking red-and-blue light in the drum kit. We ordered authentic margaritas, strong ones, with real squeezed lime juice. It was a rite of passage for me. One of Vaughn's graduate students told me that some years before at this establishment a dark-haired, rather inebriated beauty had come over, leaned on Vaughn, and whispered, "Do not be afraid, señor."

I got up to check out at closer range the guitarrone, that swollen Mexican guitar that serves as a bass. A spindly man with a mustache, Stetson hat, and gold teeth walked over to me. He had a Colt .45 (the pistol, not the malt liquor) shoved down in the waist of his pants. "Señor, you like the music?" he whispered.

A huge wave of tequila vapor engulfed me. I smiled and politely nodded.

The man's face turned sullen and threatening. "You better like it," he said compellingly.

Then a monstrous hand grabbed the mustached man by his brittle shoulder and slammed him against the wall. "Carlos, enough," this huge guy said, almost with affection. My protector turned and smiled, looking down at me from a great height. He was remarkably clean-cut and well groomed, with greased-down black hair. He stank of perfume. "How would you like to go for a walk in the lovely plaza?" he asked with even more affection.

At that moment my mates were around me and our unit made a wedge through the crowd near the bandstand. In seconds we were out in the plaza under Neil Armstrong's moon. "This may be Mike's first and last night in Santa Fe," Vaughn laughed.

As we walked back to our motel through clouds of gnats swarming around the streetlights, my head was spinning from the tequila, making it hard to get the moon in focus. I thought about that other scientific expedition, one with alloy-coated spacecraft and onboard computers. There were probably no hemp ropes or pick-axes on that expedition. Then, I thought, why not? Those astronauts were probably chunking out rock samples with geologic hammers not unlike our own. Some scientific tools have a universal application. I wondered when a paleontologist might join one of those space missions to places, like Mars, where any evidence of life on a once-hospitable planet might best be found in the rocks.

7

. . .

The Paleontological Chain Gang

"The drive from now until the end of the summer is spectacular," Vaughn announced as we headed out the next morning from Santa Fe. He was almost entirely right, and the first part of that drive was just about the best part. We threaded our way north through Rio Arriba County, one of the poorest and most stunningly beautiful regions in the United States. The sky was a dirty shroud of clouds pierced by streams of light that burned the red crests and ridges of Permian rock. The elevation was more than a mile, so the hills were sprinkled with healthy juniper bushes and piñon pines. I remembered those trips when I had visited the juniper-clad Coconino Plateau in Arizona with my parents. An old ragged-edged town of adobe looked like chips of beige chert scattered below a red Permian hill spiked with large crosses that marked Catholic grave sites. The place looked tough—the very inspiration for a spaghetti western. Indeed one of Vaughn's grad students said that in this very town he'd seen a cowboy in shit-stained sheepskin chaps and a sombrero the size of an umbrella draw his gun and shoot an old dog loping down the street.

In the late morning, Vaughn stopped at a bend in the highway and to our surprise beckoned us to start crawling the slopes of some undistinguished outcrops. "Arroyo de Agua," he announced cryptically. He started picking up some small brownish gray rocks that looked like stony turds. "Aha, got one," he said. They were indeed fossilized, defumigated turds, possibly ones left by Permian beasts some 270 million years ago. Vaughn reminded us how these lowly objects, called coprolites (from *kopros*, feces + *lithos*, stone) by paleontologists, revealed the bones of digested prey items, like small Permian amphibians.

"I wish I could get someone to work on these for a dissertation. They're dammed interesting." As Jon Kastendiek and I collected our own coprolites we agreed that this would not be our specialty. Nonetheless, there is a small cohort of coprolite specialists—coprolitologists—who find out fascinating things about prehistoric animals and their eating habits by studying these items. They are also fond of saying disarming things to people they first meet at scientific meetings, such as "My research is shit." Research on coprolites can be tedious. The fossils have to be sliced into small sections with an ultra-hard diamond-bladed saw or subjected to a rock grinder to make peels that one then embeds in resin and studies under a microscope. In this way a five-inch lump of stony turd is transformed into a thousand sections, each no thicker than an onion skin.

For most of the twentieth century these specialists were compelled to make painstaking drawings of each section of the coprolite, then skillfully re-create the three-dimensional skeleton of a tiny digested amphibian or other fossil in a wax model. Nowadays, both coprolites and fossil eggs containing embryos can be CAT-scanned. "Cathode axial tomography" is a process using lasers to make an internal image map of the object being studied. The CAT scanner is conscripted by doctors hunting for brain tumors, aneurysms, and any number of other life-threatening problems. The same kind of machinery can be used on fossils, although these CAT scanners are usually of the industrial-strength variety used for ana-

Coelophysis

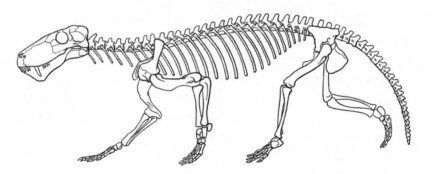

A Permian "mammal-like reptile," *Lycaenops*

lyzing metals and other solid materials—with enough energy to kill a human. The scanner makes thousands of virtual "thin sections," often clearly distinguishing the bone from the surrounding matrix. An adroit software program and a powerful computer can then reassemble these sections into a three-dimensional digital image. The result can be a stunningly vivid portrait of the original fossil, often as good as or better than if the fossil embryo or tiny digested skeleton were prepared for months by the most expert hand and sharpest needle. This is an area where technology has greatly enhanced paleontological research.

In 1969, however, I could not imagine such high-tech tools. I could only think of a graduate student back in a dark closet of a lab at UCLA patiently grinding away at lumps of petrified feces over months and months. I wanted no part of that.

We soon left the Arroyo de Agua and kept bearing north until we saw a set of magnificent pink-colored sandstones, heavily weathered into castle parapets. The color and texture of the rock was very familiar; it showed that we had crossed the line from the Permian into the next younger time unit, the Triassic, the time when dinosaurs and mammals first appeared on earth. We stopped at Ghost Ranch, where great heaps of skeletons of one of the earliest dinosaurs had been excavated. This form, *Coelophysis*, was about ten feet long and had a long skinny neck to match its tail and a small, pointed head. Its short forelimbs and long hind limbs indicated that it stood up on its hind legs. It was bipedal, in other words, like *Tyrannosaurus* and like us. I was amazed at the bounty of skeletons all exposed in the rock at the foot of the cliff. This was a La Brea–like heap of bone, except these animals were 220 million years older and more alien. I wondered with some envy if we had any chance of breaking into such a mother lode in this field season. The quarry had been worked for years by Ned Colbert, a famous paleontologist at the American Museum of Natural History in New York, the home of the world's greatest collection of dinosaurs and of a long-active and respected research program in paleontology. Vaughn remarked that these innocuous-looking creatures were cannibals. We laughed in disbelief, but he showed us a diagram of a skeleton from a specimen in the collections in New York, which outlined the ghostly image of a young *Coelophysis* bone mash in the rib cavity of an adult, about where the stomach and intestines should be located. We talked a bit about this curious situation, and we thought of those male lions known on occasion to eat the cubs in their own pride.

These cannibalistic dinosaurs and other prehistoric beasts from Ghost Ranch lived on the younger side of a great and terrible event that separated them from *Dimetrodon* and other denizens of Permian time. At 250 million years ago, 30 million years before *Coelophysis* thrived, the greatest mass extinction event of all time marks the boundary between the Permian and the Triassic. At that time, one single gigantic landmass, Pangea, embraced virtually all of the earth's present-day continents. Most of North America west of present-day Colorado and New Mexico was under a vast ocean. Marine creatures in the Permian-Triassic rock sequence from Alaska, the Italian Alps, and many other places record the devastation of the reefs and other marine fauna—over 95 percent of all Permian marine species did not live to see the Triassic. In addition, several major groups that included these species as members were virtually wiped out. These include the brachiopods (lampshells), bryozoans (colonial moss animals), corals, crinoids (sea lilies), shelled ammonoids (fossil relatives of the living nautilus and, more distantly, of the octopus and the squid), and trilobites. It took millions of years for these organisms to rebound and recolonize the ocean.

By odd contrast, several groups suffered some species loss but as a whole came through the Permian extinction with healthy diversity. These resilient sea creatures included gastropods (snails), bivalves (clams), and crustaceans, such as the early ancestors of lobsters and crabs. On land the Permian extinction was only a little less pronounced, affecting between 75 and 85 percent of the species. The traumatic change in the land animals is especially well recorded in South Africa, where the boundary between the Permian and Triassic is exposed in road cuts in places like Lootsberg Pass. Here, Permian rocks preserve an impressive array of synapsid species of a more specialized variety than *Dimetrodon*. Mixed among the synapsids are a few specimens of an ugly, rather stumpy beast

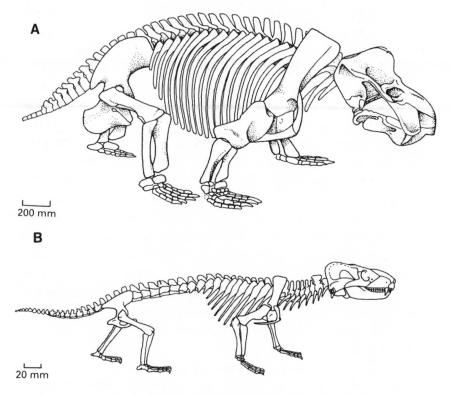

A

200 mm

B

20 mm

Two Early Triassic animals: (A) the dicynodont *Kannemeyeria*, a
close relative of *Lystrosaurus*, and (B) the cynodont *Thrynaxodon*

with a flat face and large protruding tusks called *Lystrosaurus*. Just
above the Permian rock layers, in the lower Triassic sequence, the
diversity of synapsid species decreases drastically. But *Lystrosaurus*
and a few others have their day—they greatly increase in abundance, even though the diversity of synapsid species remains greatly
impoverished.

A Permian catastrophe is also evident in the case of land plants.
Paleontologists scaling the steep slopes of the Italian Alps have sampled the rocks for microscopic fossils—pollen grains from plants
and spores from fungi (molds and mushrooms) that feed on decaying vegetation. Pollen is abundant and dense in the lower part of
the rock sequence marking Late Permian time, but it is very scarce
in the bottommost layers of Triassic rock. In fact, the Permian-

Triassic boundary is marked by a huge explosion of millions of fungal strands in small golf ball–sized rocks. The researchers claim this indicates a fungal feeding frenzy on rotten wood from trees all over the world that were destroyed during the great extinction.

The Permian extinction is the fossil record's most powerful statement that life on this planet can be as ephemeral as a dandelion in a gale. It is the greatest of the five great extinctions, of which it is the middle one. The last of the five was the event that wiped out the nonavian (nonbird) dinosaurs and many other terrestrial and marine species at the end of the Cretaceous, 65 million years ago. None of the other four events matches the Permian extinction for sheer devastation. For example, the Cretaceous event may have claimed "only" about 70 percent of all species. Major Cretaceous groups on land and in the sea survived into the next period, the Tertiary, despite their winnowed numbers of species; on land only the nonbird dinosaurs and some groups of mammals were snuffed out. For some reason, many other groups—such as turtles, crocodiles, certain other mammals, lizards, gars, and a few freshwater fish groups—managed to hang on into the Tertiary Period and even flourish. So the Permian extinction exceeds the intensity of all others, not only in terms of the species lost, but also in terms of the number of major evolutionary groups and ecosystems that were disrupted.

It is hard to estimate how long the Permian extinction actually lasted. Some geologists put the phase of decimation somewhere around 100,000 years, but it could have occurred over a far shorter period of time. From our disadvantageous viewpoint in the present we don't have the scientific tools to resolve precisely the timing of events that occurred a quarter of a billion years ago. Another limitation of our understanding concerns the actual cause of the Permian event. There is circumstantial evidence that at this time there was an inordinate amount of volcanic activity. There also seems to have been a very high influx of toxic levels of carbon dioxide in the oceans, which may have transformed the waters of life into a deadly bicarbonate soda. There is even some evidence of an asteroid impact. A crater recently discovered in western Australia is seventy-five miles wide and could have been created by an asteroid three miles across. As in the case of the Cretaceous extinction, such an impact could have sent enormous quantities of dust into the atmosphere, changing the climate by creating more cloud cover and possibly even blocking the sunlight vital to plants, which died away along with the struggling organisms that fed upon them.

This chain of events is not clearly documented, however. No one cause has been championed with any great success. Many Permian specialists admit that toxic oceans, volcanism, asteroid impact, and some other event may have combined to deal the death blow. In formal scientific argument such theories are said to be pluralistic, but pluralistic theories aren't very useful because they don't allow you to home in on a single elegant explanation and test it against new observations or experiments. If we said the orbits of the planets in our solar systems were due to many factors, this would not be as satisfying as the accepted theory of planetary motion, which ascribes these orbits to the gravitational pull of the sun. There are simply too many reasons suggested for the source of the Permian extinction, and fossil and geological evidence gives no indication of a clear overriding cause.

Whatever its cause or causes, the terminal Permian extinction tore a gaping hole in the earth's ecosystems. Gone were the magnificent multihued reefs, the thick conifer forests, and the dragonlike synapsids that represented the dominant, large-sized land vertebrates. It took several million years for these communities to reemerge, with many new species filling in the ecological roles vacated by their extinguished predecessors. By 220 million years ago, when the rocks were laid down in western North America in places like Ghost Ranch, the new landowners were of course some of the first dinosaurs.

The sun was low as we left Ghost Ranch and hit the long upgrades that marked the approach to the Rockies. We soon made camp in what was the first cool, actually frigid, night of the trip. I shivered in my Sears army-green sleeping bag with its Scotch plaid lining, hoping for sleep to stave off the cold. But sober reflection on our lack of success and anticipation of many more unproductive days kept me awake. When I closed my eyes, I saw my shoes plodding along through the red sand on the slopes of the Caballo

Mountains and my hands picking up the puny bone fragments of some fin-backed fossil reptile that seemed to be everywhere in New Mexico but never preserved in a resplendently complete state. Early in the field season, I had impatiently and impetuously asked Vaughn where were those really good skeletons of *Dimetrodon*—like the ones illustrated in kids' fossil books. "Central Texas," he said with his usual terseness and perhaps a bit of annoyance. Based on my firsthand field experience, *Dimetrodon* and other synapsids were already virtually extinct millions of years before they were supposed to have been snuffed out.

As I drifted into semiconsciousness, I heard the hiss of a river far below in the canyon, and I dreamed of the azure waves and the warm sands of Santa Monica Beach. How long had I been away from home? Two weeks? Four weeks? I had a hard time reconstructing near-identical days with endless excursions up canyons devoid of fossils. Vaughn had said the trip would take eight or nine weeks. So maybe we had thirty-five days, or 840 hours to go. The prospect of more days like the last ones seemed intolerable. I came back to my earlier realization, one that first emerged during those bone-dry days in southern New Mexico—field paleontology could be boring, especially when you weren't finding anything.

But even boring field paleontology could take you to nice places. The next day we kept gaining elevation, passing shimmering alpine fields dusted with blue flax, Indian paintbrush, and columbine. We tipped over Wolf Creek Pass in southern Colorado, one of the highest stretches of highway in the country. From that splendid point we could see waves of mountain summits still shrouded in snow. We were, as promised, headed for cool mountain country. Vaughn stopped us at a public campground at the eastern base of the Sangre de Cristo Mountains, about thirty miles south of the town of Salida, Colorado. There we would camp for at least two weeks and work a quarry somewhere in the scruffy foothills to the east. We were a rough-hewn bunch by then, eschewing baths for several days and subsisting on boiled beans and "salads" made from sliced cucumbers, onions, and white vinegar. We felt strange in a community filled with sauropod-sized Oldsmobiles with Oklahoma and Texas license plates, towing bullet-shaped Airstream trailers into the overstuffed camp. I looked out at the humanity from America's flatlands and got a jolt of recognition of the benefits of field paleontology. It was good to have work that got you away from normal people and planted you in forlorn, empty places, like the Caballo Mountains, with a few other eccentrics.

Despite our sudden confrontation with other people, it was worth putting up with the noisy campground, for soon we finally got a taste of what it meant to find some good fossils. The next day began inauspiciously, like all the others preceding it. We left camp and turned north on the highway toward Salida that runs along the Arkansas (locally pronounced ArKANses) River. After a few miles we turned east, crossed a cattle (or cable) guard, and drove down a dusty road that skirted a pebble-filled wash incised by a rushing creek. Then we turned a sharp right up a very steep embankment and struggled up a boulder-strewn ridge in low gear. At the top of one of these ridges we parked the Jeep and truck beneath piñon pine and descended on foot into a grass-covered ravine. Vaughn stopped enigmatically at a small outcrop of blackish shale, no more than seventy feet wide. I noticed there were small shards that looked like chips of flint strewn all around; the place was like a factory for Indian arrowheads. I also noticed that some of these pieces were particularly shiny and black.

"Bone?" I asked, holding up a fragment with a smooth surface that made it look blacker and more reflective than the surrounding rock.

"Nope—slick'n'side," Vaughn said, referring to the distinctive facet on a rock surface caused by a chisel.

Undaunted, I picked up a bullet-shaped piece that looked like a fragment of a tiny limb and held it up to the sunlight. "Bone?" I asked again.

"Sure is," Vaughn nodded and instantly squatted down, brushing away a lot of dusty gray-black rock with his forearm. Then he walked over to the heavy burlap sack, grabbed a five-pound hand sledge and a few cold chisels, returned to his former station, and started banging away. Our reaction time was a bit slow. "This is my lucky spot, but there's bone all through here." He impatiently waved his hand toward the outcrop, as if the rocks were waiting for us to start pounding.

This ritual inaugurated my first prolonged experience at professional fossil quarrying. At first the work seemed rather haphazard and not very different from chopping out spiral cones of fossil snails in the Santa Monica Mountains, breaking nodules open for trilobites in a northern Wisconsin quarry, or digging out geodes full of quartz crystals with Russ, Mick, and my brother in the Potato Patch west of Yuma. But it was clear there was a subtle technique in hitting the surface of the rock just right to crack it. The chisel had to bite into a good crack without shattering the slab. Vaughn patiently showed us how to lay a chisel in at the right angle and to bury it with only a few strategic hits. But more important than a well-placed chisel and a productive swing of the hammer was a good eye to guide your quarrying. The first glint of decent bone—a small end of a finger bone, or a rib, or a tooth—required a careful rerouting of the chiseling that skirted the fossiliferous slab and left it intact. This was often impossible, because the same crack that loosened the slab also split the emplaced bone. But Vaughn wasn't upset about this first accident. He immediately applied a glue mixture thinned with acetone so it would dry and harden the bone without coating it too thickly. Globs of glue were an annoyance, because they would require painstaking removal with a needle later in the lab, in order to expose details on the bone surface. After a few minutes of drying, he would very gingerly start making a trench with the chisel a bit farther away from the bone. Thus, finding more bones meant more work, as trenching in ever-widening circles meant larger blocks of bone-filled rock.

At first this intensity of labor wasn't much of a problem. We weren't finding much bone at all. Then we heard Vaughn murmur. I turned around to see a set of delicate, flattened, nubbinlike objects that extended into a series of four or five segmented finger bones. The whole hand was no larger than a quarter, but the black shiny bone formed the extremity of a beautiful little vertebrate fossil. As Vaughn had predicted when he first invited me to the field several months before, the animals in this quarry near the Sangre de

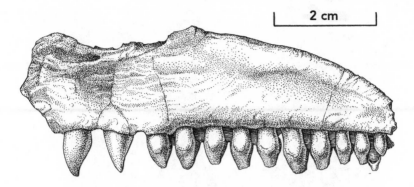

The upper jaw and teeth of *Desmatodon*, from Vaughn's Interval 300 Quarry

Cristos were mostly rather small, delicate creatures. The quarry did contain bone shards of some bigger animals—synapsids, such as a primitive species of *Edaphosaurus*, and largish sluglike amphibians known as labyrinthodonts, in reference to their complex, labyrinthine teeth. There were also scales of fishes and bits of calcified fossilized cartilage of a very weird group of sharks called xenacanths. But the real prizes within the quarry were very small amphibians belonging to two groups, the aistopods and the microsaurs. These were known from skeletons no longer than seven inches (18 cm), smaller than those molting alligator lizards in the backyard of my parents' house. They were important fossils because they represented branches of the evolutionary tree between those of amphibians and reptiles. These intermediate forms present difficult problems. Their anatomical details are a complex mixture of primitive and more specialized features, or amphibian and reptile features. But paleontologists love such forms because they demonstrate that fossils have much to tell us about evolutionary history that cannot be learned from studying only living organisms. They document the steps in the emergence of the reptile groups, or the transition of vertebrates from water to land, or the origins of flying forms. In regard to the last, think how impoverished our understanding of the evolution of birds would be without the fossils of *Archaeopteryx* from 145-million-year-old rocks in Europe, which allowed for the riveting disclosure that the earliest birds were simply small predaceous dinosaurs with feathers.

A coal forest of Pennsylvanian age, as depicted in an 1892 book by Ernst Heinrich Haeckel. The giant reedlike plant on the left with leaf whorls is *Calamites*

The little gems here in Colorado—the aistopods and microsaurs—that were most coveted among the fossils from this small outcrop were mixed in with large amphibians and fish in a bone assemblage that was very old—about 300 million years old. This time represents the Pennsylvanian Period, which predates the Permian. The Pennsylvanian is so named because a rich, coal-laden rock sequence of this age is well exposed in the hills of Pennsylvania as well as in adjacent eastern states. These rocks are stuffed with plant remains that have been compressed to form rich coal seams. The Pennsylvanian sequence in North America and elsewhere thus preserves a world of steaming bogs and swamps set under the shade of cycads, ferns, and other primitive land plants. Dragonflies with two-foot wingspans glided over the bogs, and fat labyrinthodonts shared fallen logs with delicate little aistopods and microsaurs. It was a North America very different from the dry deserts of the Early Permian where *Dimetrodon* snarled. The locality we were working was in fact very significant in another way. It was the only known assemblage of Late Pennsylvanian age west of some well-known localities in eastern Kansas. It was thus important in giving us a better sense of where creatures of this age lived and how they might have differed from place to place.

As we carefully glued the small bones of this 300-million-year-old amphibian, I wondered how Vaughn happened to find such an obscure locality only seventy feet wide in the middle of a wilderness of grassy hills and piñon. Vaughn was quick to acknowledge that an unpublished master's thesis on the stratigraphy of the area by one Walter Pierce put him on the trail. This measly outcrop was part of "Pierce Unit V" in the lower part of the Sangre de Cristo Formation. Although it was exposed here only to a thickness of a few feet, the total thickness of the rock formation in this region was about 8,800 feet. The fossil-rich shales dipped into the ground at a steep angle, so one could surmise that most of this rock unit was well buried. Pierce had mapped this little outcrop in its estimated stratigraphic position relative to the total thickness of the Sangre de Cristo Formation as "Interval 300." Vaughn accordingly applied the undramatic name "Interval 300 Quarry" to this fascinating pocket of ancient amphibians.

The next step in quarrying occurs when one has a sense that the bone or the skeleton goes no farther into the rock. At this stage, we stabilized the exposed bone by slathering on a bit of water-soaked toilet paper and letting it dry. In working out this little amphibian we found that it was not very complete. That was a disappointment, but still it was an exhilarating find. We then capped the top of the slab, no bigger than a pillow, with a couple of layers of burlap soaked in plaster. In the hot gully of the mid-afternoon, this was the fun part. The plaster was cool to the touch, and we used this procedure as an excuse to douse each other with water carried from the icy stream nearby. When the cap was dry we started making a deep trench around it, three people applying cold chisels in strategic directions and angles. This progressed until the cap was substantially undercut, so it looked like a white toadstool. Then came the tricky bit. At the moment of truth, Vaughn drove a long chisel nearly through the pedestal, loosening the rock from its moorings. All hands clutched the edges of the plaster block. We then heaved the specimen over onto its plaster shell, praying that the shale wouldn't simply crumble away and spill out. The plaster block flipped over beautifully, with nary a loose piece of shale spilling out. We then carefully trimmed the uneven underbelly of the cast with chisels and big knives and slathered it over with more plaster-soaked burlap and a thick frosting of extra plaster. In truth, there was less drama than might be expected in overturning this cast. Vaughn was methodical and exacting in his calculations. This block, even with its thick coating of plaster, weighed only a hundred pounds. I wondered what reserves of strength and engineering were required to flip half-ton plaster monoliths that encased the limb bones of dinosaurs.

Like other Pennsylvanian rock sequences, the Sangre de Cristo Formation had dense remains of plants. Although only a few plant fossils were found in the Interval 300 Quarry, Vaughn took us to some nearby outcrops that had loads of plants. Here we separated slabs with impressions of ferns and casts of the bumpy stems of ancient *Calamites*, a horsetail that grew as high as fifty feet. It was fairly easy to visualize the thick tree fern and horsetail forest around a 300-million-year-old bog in what is now Colorado. Vaughn remarked on the absence of bone in these plant quarries. "Bone often gets etched away by decaying plant material, it becomes part of the compost heap. Good fossil plant localities are often bad fossil bone localities."

After a day of quarrying we were eager to get some circulation back into our limbs. The cooler evening encouraged a hike up the steep trails to the snowfields several thousand feet above our campground. The summits of the Sangre de Cristos are also of interest to paleontologists. The skyline ridge is often made of a swirl of maroon sandstone. Because of the uplift, folding, and overturning of layers of rocks that came with the violent birth of the Rockies, these high sediments were actually much older than the Pennsylvanian-aged Sangre de Cristo Formation in the valley far below. Maroon boulders weathered out and came to rest on the side of a snow slope or the muddy shore of a sapphire alpine lake. These boulders also had little flecks of blue-white bones, skeletal bits of some of the first fishes, which populated the oceans some 350 to 400 million years ago.

For several days we broke rocks at Interval 300 Quarry, telling rude jokes, partaking in plaster fights, and hauling out some decent fossils. Our hands were swollen and blistered from swinging sledges all day, and our backs were sore from cradling heavy plaster blocks as we climbed the steep slope back to the vehicles. The work was hard. But I liked the routine, I felt useful; we were a chain gang in the service of science.

8

■ ■ ■

Red Rocks

As the days passed at Interval 300 Quarry, our feelings of accomplishment wilted and the work got harder. Bones were thinning out, because we didn't have much of a working face to quarry. The bone-bearing shale layer plunged deep down and away from us, covered by several feet of barren Sangre de Cristo Formation. This required even more energetic chopping and shoveling. My hands were covered in crocodile skin; I wondered if I would ever again be able to finger those intricate chords up the neck of my guitar. One day, Vaughn quietly set down his hammer and chisel. He stood up and stared intensely at this uncooperative overburden, tilting his head up and down as he scanned the profile of its stony face. I thought for a moment that we were about to get ready for some really stupendous labor in our little Pennsylvanian coal mine. But Vaughn instead said, "It's time to go, boys. We'll come back here next year—with dynamite." He wasn't kidding. There was plenty of precedent for setting a well-placed charge to blast out rock and still have some undamaged bones to show for it. Such an operation would require National Forest Service permission, though, even in those less restrictive times, and we weren't equipped for the demolition procedure in any case.

It did not take us long to pack up camp and wave goodbye to the Texans and other "flatlanders" sitting in lawn chairs by their Airstreams. When we stopped in Salida, Vaughn mentioned that a dispatch he'd picked up from the post office indicated that his graduate students would not be joining us. I was relieved, not just because this meant no diabolical pranks. By this time I had grown used to living with this obsessive paleontologist and our small band of tyros. My peers and I were unabashed in our ignorance, unhesitant to ask the most basic questions about geology or fossils or places, questions which Vaughn for the most part patiently answered. We relaxed in our equally low status, as there was no reason to try to excel in our field performance under the circumstances. Vaughn demanded hard work—he had harsh words for field hands who had absolutely no interest in or affection for fossil hunting—but he did not expect consistent excellence in all duties. He did not mind our unconcealed boredom when pounding away in a sizzling rock quarry, as long as we sometimes showed a genuine connection with the work and with the grand landscapes of the West. Too many complaints about discomfort, dirt, or bad food were not tolerated. He often remarked, "I like intelligent slobs, not fastidious idiots."

Vaughn also gave us time to ourselves; I would often hike alone up to the snowfields above our campground and sit on a 400-million-year-old boulder of maroon sandstone sprinkled with fossil fish bones, gazing out over the rolling, tawny hills under the maroon evening sky over eastern Colorado and distant Kansas. No one expected me to always be around and always be the "team player." Vaughn knew the feeling of freedom in the hills, and he respected our own desire for that freedom.

The route we took out of Salida was like an old mountain man's trail. Vaughn, to our delight, liked to negotiate the steepest, most deserted, and most scenic of mountain passes himself. He guided us to the mining town of Telluride. The place was magnificently rustic. Saloons and stores were made of wood planks weathered into silver slivers; the siding resembled the back of an old grizzly bear. The town graveyard was perched on a hill so steep it looked as if the corpses must have been buried standing up. It was impossible to forecast from our visit the kind of growth and commercialization that would one day make Telluride a posh ski destination for the Hollywood glitterati. From the hilltop cemetery we could see the slopes where the future ski trails would be cut. Without a winter coating of snow they were impressive ruts that descended for a couple of thousand feet through deep red Permian rock.

Traveling toward Monument Valley from Telluride, we could see the easternmost snowcapped range in Utah, the La Sal Mountains, poking up above a brilliant red sequence of Pennsylvanian, Permian, and Triassic rocks, layers swirled into giant ridges and folds that looked like the backbones of huge dinosaurs. The amount of rock—the pure extent of red from one horizon to the other—was astounding. It was as if we were on Mars. Every crack and wrinkle of this terrain lay fully exposed—ridges, tilted strata, folds, overturned beds, deep canyon walls cut by unfamiliar rivers, faults that offset similar rocks, towering spires and monuments. The landscape wasn't so deep or intricately weathered as the Grand Canyon, but it went on for hundreds of miles, and it was flat enough to allow us to drift among its stone troughs and swells. This was the most naked rock I had ever seen, as if the Grand Canyon had been unfolded and laid flat like a blanket on the earth's imperceptibly curved surface. It was, indeed, the great red heart of the Colorado Plateau.

We spent fourteen beguiling, beautiful, and mostly unproductive days in the red rock country. I walked long hours around giant columns of sandstones looking for ancient amphibian skeletons like *Tseajaia*, the animal so stunningly represented by a skeleton in Vaughn's lab. We greenhorns were all skunked; only Vaughn seemed to have an eye and perhaps a well-seasoned obsession strong enough to extract some decent skeletal fragments from two rock units, the lower and older Halgaito Shale and the higher and younger Organ Rock Shale, the alleged source of *Tseajaia*.

With rocks so obstinately unproductive, we wondered what jus-

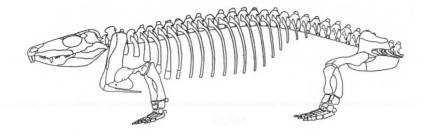

Tseajaia

tified our long and fruitless days in Monument Valley. After our respectable findings in the Sangre de Cristo Formation in Colorado,
why didn't we point our vehicles east and south to the land of the
flatlanders and scour the washes of north-central Texas? After all, the
wide belt of grazing territory stretching from San Angelo north to
Wichita Falls could claim the richest assortment of *Dimetrodon* and
other creatures from the early Permian of any region in North
America. Vaughn was somewhat casual with his explanation, claiming that north-central Texas was ugly and Monument Valley was not.
He even remarked that he had to go to Monument Valley on pilgrimage at least once a summer, for here John Ford filmed his classic John Wayne westerns like *The Searchers*. As expected, famous
quotations from Mr. Wayne were offered with even more frequency.

But his evasiveness belied the solidly scientific reason for working the red rocks of Utah. For many years there had been a popular theory that early Permian vertebrates from the Four Corners
region were very different from those that lived at the same time in
north-central Texas. Moreover, it was hypothesized that the reason
for this apparent difference was that an inland sea separated the two
land regions. Although there were extensive marine Permian rocks
in western North America, these were largely confined to southeastern New Mexico and southern Texas. In the Four Corners region there was also evidence of an extensive seaway, largely to the
north and west of present-day Monument Valley. This western seaway was represented by the Cedar Mesa Sandstone, which intruded
in a wedge between the older Halgaito Shale and the younger
Organ Rock Shale. What this interaction indicated was a shoreline that had shifted through several million years. There was no

strong geological evidence that either of these seaways extensively flooded areas between Utah and Texas. Nonetheless, paleontologists thought that such a sea might have existed.

Vaughn's famous professor, Alfred Sherwood Romer, had claimed that Texas was full of water-dwelling species—like the early lobed-finned fishes that were related to the first land vertebrates, the big, sluglike labyrinthodont amphibian *Seymouria*, and a more delicate small salamanderlike amphibian, *Diplocaulus*. In the same paper, published in 1960, Romer emphasized the great abundance of *Dimetrodon* in Texas; farther west, instead of *Dimetrodon*, he noted only the presence of a less flamboyant *Dimetrodon* relative called *Sphenacodon*, a form with very short spines and thus an animal for which one could not reconstruct the famous *Dimetrodon* sail fin.

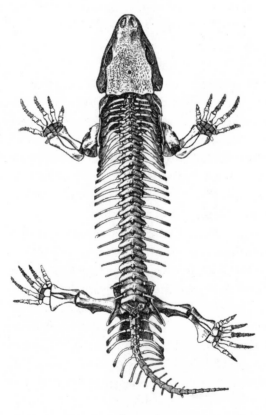

Seymouria

Vaughn had already disproved part of his professor's argument by finding scrappy evidence of *Dimetrodon* in southeastern Utah. He had also recovered remains of lobed-finned fishes, *Diplocaulus* and *Seymouria*, in southeastern Utah, even though these fossils were apparently absent in the northern New Mexico localities. In a paper published in 1966, Vaughn had attributed these differences to the fact that the animal communities, or faunas, of Utah and Texas shared basically the same kind of habitat—a broad floodplain or river delta near the margin of a shallow sea. Thus, both regions shared an abundance of creatures particularly suited for life near water. By contrast, the northern New Mexico localities, he suggested in the same paper, revealed faunas that represented highlands farther removed from the inland seas. Vaughn had demonstrated that patient and resolute prospecting in the Four Corners revealed the presence of additional Permian animals. Moreover, this simple comparison of samples led to an interesting geographic pattern: the same fauna in the lowland areas of Texas and Utah, and a different fauna in the highlands of the intermediate area of north-central New Mexico. Importantly, the paleontological story was now more consistent with the geological story: there was no reason to propose a sea barrier between the Permian landscapes exposed in Texas and Utah.

Our puny pickings over the last days of the 1969 season bolstered Vaughn's geographic story for the animals of the Permian, but failed to do much more in the way of new animals or better bones. It had been a long summer, and I admit to counting the minutes until my return to L.A. Vaughn seemed sympathetic— maybe empathic—and we relieved the frustrating bouts of prospecting with some pleasant diversions. We would stop for lunch at the Desert Queen Café in Mexican Hat, where an extremely massive woman, who, we speculated, was indeed the Desert Queen, would give us a friendly greeting and offer us the eternal special of the day—pea soup and chicken-fried steak. We beat the heat by swimming in one of the graceful bends of the Green River, spent time sifting through piles of Navajo saddle blankets at the Old Trading Post at Tseatsose Wash, and rummaged through the public library in Blanding, Utah, where there seemed to be no books on

Russian subjects, not even Dostoevsky's *Crime and Punishment*, because anything Russian smacked of communism.

These extracurricular activities broke the routine of desperate prospecting until the planned end. No fabulous fossil discoveries compelled Vaughn to extend his field schedule. There was also an ominous *click-clack* from deep within the bowels of our jeep. "Sounds like a broken tooth on our low gear. Time to limp back to the UCLA motor pool," Vaughn pronounced.

We finally arrived in Los Angeles on a hot late August day. I was thrilled to be back. As I drove the truck down Sunset Boulevard (I had learned to drive tolerably by this time), a Hendrix guitar riff growled out of the radio. A parade of women in bikinis were gathered at a corner, waiting for the bus to the beach. We were passed by sleek white Corvettes and canary yellow Porsches. I thought, What a weird and wonderful place this is. I was eager to rejoin my band and play some music. I thought I had purged my soul of any desire to tramp, most of the time in desperation, over dry ground. The red rocks were great, but the adventure was over. Time to be back in L.A. town.

9

. . .

Back to the Bones

It is disturbing, as I was to relearn vividly in later years, that the contentment felt upon returning home from the hinterlands can be so ephemeral. Despite the respectable findings in the Sangre de Cristo Mountains of Colorado, the field season of 1969 had hardly been one of great triumph. Yet I couldn't abandon the connection with that long, hot summer. I couldn't resist stopping in to bother Vaughn as he picked away at the motley assortment of fossils from Monument Valley or the more impressive specimens from Interval 300 Quarry. As he had predicted, the 1969 summer field season turned out to be 49 percent anticipation, 2 percent success, and 49 percent recollection. That recollection as well as my growing interest with paleontology blurred the memories of discomforts on the Jornada del Muerto. I accepted Vaughn's invitation to join him again in the field in the summer of 1970. Ironically, despite my growing disenchantment with a career in music, the band actually had some jobs and sparked some interest from a major recording studio. So I rendezvoused with Vaughn and his crew later in the season at Interval 300 Quarry. This meeting was facilitated by a friend, Craig, who had just bought a new four-wheel-drive Ford Bronco. Craig offered to drive me to Vaughn's camp and then intended to split off, roaming the Rockies in solitude.

By this time I knew enough of the West to be a useful guide, and we took our sweet time getting to work, indulging in any excursions that properly challenged the Bronco or our new mountain boots. We climbed Mount Ellen, northeast of the Escalante canyon country in Utah. This was the last of the major peaks in the interior United States to be scaled, and the reason is evident from its extreme isolation. Mount Ellen stands like a watchtower over an endless desert. From its lonely summit we peered east to where the confluence of the Green and Colorado rivers creates an enormous Y-shaped wrinkle of canyons in the earth's crust. The experience was not, however, entirely aesthetic. As an act of celebration we decided to take the risk of sleeping on the brittle, razor-backed ridge of the summit. We secured our sleeping bags to the rocks lest they drift off the edge of a precipice that dropped more than 2,000 feet to a tundra carpeted in columbine. At 2:00 a.m. we were awakened by an enormous storm rolling in on us from the confluence. This required hasty retreat in a night so dark we prayed for lightning to show us the way down. In munificent response, the heavens opened and the lightning started barraging us like a cannonade. As we traversed a steep scree slope soaked by icy rain, I slipped and rolled for thirty feet until I stopped on a ledge. I lay there disoriented for some seconds. Craig shouted down to me. "Nothing broken," I cried, and rose to grab the handhold above. We staggered down the drenched cliffs and ridges for hours. As the sun hit Mount Ellen's summit we could finally see the Bronco parked on a saddle of bare rock a thousand feet below.

Next I took Craig to Elephant Hill in Canyonlands National Park, where we nearly lost the Bronco down a series of steps in the sandstone known as the Elephant's Staircase. We then drove on to Telluride, made a much more civilized climb of nearby Mount Sneffels, one of the highest peaks in the San Juan Mountains, and

finally got to Vaughn's base of operations in the crowded campground at the foot of the Sangre de Cristos. "Long time getting
here, you missed all the work," one of the crew said uncharitably.
After bidding Craig farewell, I joined the others and pitched in
with extra effort at the quarry, as if to compensate for my tardiness.

This year the quarry operation was a big production. Everybody
wielded extra-large sledgehammers; Vaughn was trying to maxi-
mize the exposure of the rock that covered the fossil-bearing layer
before he blasted the overburden out of existence. One day, when
someone was pounding a small metal wedge to secure a hammer-
head, a piece of shrapnel flew off and penetrated my right arm. A
fountain of blood arched magnificently from my arm to a spot several feet away, nearly drenching Vaughn. "It's a radial artery!" I
exclaimed, as if to prove I had learned something in Vaughn's comparative anatomy course. At that time, we applied a remedy now almost universally rejected by climbers and hikers in the wilderness.
Pressing the wound with rags, Vaughn twisted a bandanna into a
tourniquet farther up the arm. As he took me to the Salida hospital, one of the grad students in the back of the car, a diabetic, started
to faint from insulin shock. Vaughn needed an extra shot of tequila
that evening.

At last we were ready for the big event. Vaughn extended a fuse
line for nearly a hundred yards up the creek, over boulders, and
through the trees. It reminded me of the dynamite line Alec Guin-
ness discovered leading to his precious bridge on the River Kwai.
Vaughn stood at the charger, which looked exactly like the ones
rendered in a Daffy Duck cartoon, and pushed the handle with
both hands—*Kaboom!* I was actually a little disappointed with the
muffled blast, but when we returned to Interval 300 Quarry it was
a site of pure chaos and devastation. Vaughn was of course panicked
that he had inadvertently destroyed his precious vein of fossils.
After hours of removing shattered rock we found to our relief that
only a small chunk of the bone-bearing shale had been damaged.
The rewards from that risky operation were indeed dramatic. One
day, while we quarried a clean and productive rock surface, part of
a beautiful skeleton of a microsaur amphibian popped out. We
made a very hefty jacket for this specimen and pulled it up the

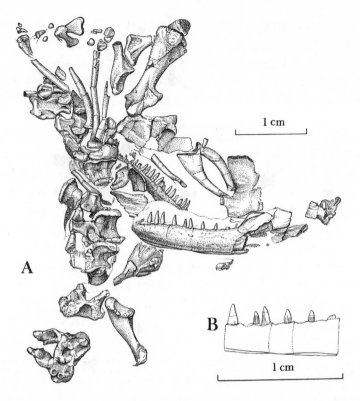

The disarticulated skeleton (A) of *Trihecaton,* an ancient Pennsylvanian-aged amphibian. A detail of the lower jaw is shown in B

steep hill with the help of a motorized winch on the front bumper of the jeep. Back in the lab, the specimen turned out to be a tiny and delicate skeleton, with a skull no longer than one inch and a string of vertebrae less than seven inches long. The skeleton belonged to a new and important amphibian, which Vaughn named *Trihecaton howardinus* in a paper published in 1972.

The fieldwork in Monument Valley and parts nearby was again less productive. At that time I did not assiduously take field notes, so I don't have much to help me recall the events that occurred now more than three decades ago. What I do remember most about that time in Utah was a nightmare. We were tracking the Permian exposures near the Goosenecks, the great loops of cliffs that border the Green River. Vaughn led us into a deep defile, Johns Canyon, named for two brothers who once had a small ranch there. According to grisly legend, one brother had killed the other and thrown his head into the deep gully next to the ranch. We, of course, had to camp next to that haunted edifice. I slept in the deserted dirt road to keep some open ground around me, as nocturnal sidewinders in Johns Canyon were very active. In a dream I saw a moon and an empty dirt road, and a very large woman, indeed the Desert Queen, lumbering toward me, smiling sardonically. In one hand she swung an ax, in the other she clutched the severed head of poor Mr. Johns by his bristly, bloodstained hair.

Johns Canyon was virtually the last stop on the return to L.A. Once home, I realized I had more than a casual curiosity about the reasons fossils are hunted in the first place. I signed up for graduate seminars in paleontology even though I was still not a fully fledged graduate student myself. There we debated the intricacies of the evolution of vertebral structure from fish through amphibians to reptiles and the geographic problems of the Permian West. It was my first experience of real scientific dialogue, argumentative give-and-take, and honest expressions of ignorance. In this arena, even Vaughn and his most seasoned doctoral students frequently answered questions with the response "I don't know."

Such absence of hubris was well justified. Time and time again we reviewed cases where theories had seemed so strong, only to be toppled with one good insight, one more observation, or one more fossil. It was amazing to think that the most respected geologists and paleontologists had believed only a few years before that the continents of the earth were isolated and immobile over the ages. These experts dismissed the theory of continental drift as just a wacky notion conjured up in the 1920s by an eccentric meteorologist named Alfred Wegener. The confirmation of continental drift through the theory of plate tectonics was a long-overdue vindication of Wegener and a brutal shock to the cognoscenti who had ridiculed him.

The refutation of some theories, such as Vaughn's rejection of Romer's proposal that Permian vertebrates from Texas and Utah were kept from interchange by a broad inland sea, were of less global significance, but they contributed their own pleasing nuance to the history of life. The process of theory and test had a particular elusive rhythm, which you know you've got when you get it, like the first time I could play all the chords of "House of the Rising Sun," my first song on the guitar. The probing, the mystery, and the debate necessary to attain at least some temporary level of understanding became exciting to me.

So this is what science is supposed to be. Not bad. It struck me that all that roaming around in wild places, those endless days of walking up beautiful canyons, those campfire dinners with sinewy *conejo*, those spectacular drives through red rocks, those tequila bars, those girls in the plaza at T or C, were all part of a scientific adventure. And some people, like Vaughn, made a living at it. He seemed much better off than the gray-skinned, white-haired organic chemistry professor, who through either misanthropy or an inability to communicate his profound knowledge of the subject could not construct an exam with a higher average grade than 58 percent. (During a midterm exam, a German shepherd entered the chemistry lecture hall filled with 150 miserable students, and someone yelled out, "Get that dog out of here, he'll raise the curve!")

Vaughn's courses were not at all easy, but he had the rare gift of a good teacher. He gave you the opportunity to appreciate the story of evolution, whether translated from comparisons of anatomy in living creatures or read in fossils in the rock. You could reject that story as uninteresting, but at least you had a fair chance to get the straight story in a clear and stimulating style. And who could deny the advantage of testing your theories by searching for fossils in a rosy canyon in Monument Valley rather than in some sickly green, fluorescent-lit chemistry laboratory? I couldn't shake the pleasures of ambience, discovery, and even ennui that came with roaming the Colorado Plateau in search of old bones. Slowly, half-consciously, I was becoming a paleontologist.

10

■ ■ ■

The Pits

My distractions with school matters and summer fieldwork had cut into my earning power as a musician, and I needed some extra cash to contribute to the rent of a small house filled with musicians and streams of visitors, including groupies and their dogs. At the age of twenty, I was clearly at a low financial point; I was hard up, and to make matters worse, I needed flexible hours and employers with a tolerance for long hair. In desperation I answered an ad for a sort of zookeeper, only one for domestic cats. I rationalized that it was, after all, the care of animals, and I wasn't dainty about nature's products, like cat shit. Besides, it only meant three or four two-hour stints a week at a surprisingly good salary. It did not take long to find out why the pay was so generous. I shall not go into the details of that miserable job (the cages for Siamese cats were particularly disgusting). Suffice it to say I carried out this work in strange company. The supervisor was a chain-smoking woman with a rasping cough who looked a bit like Janis Joplin and spent virtually all her time watching game shows on television. My other cat-care comrade was a night watchman who was trying to make inroads into Hollywood by carving alien space monsters the height of basketball players. One of his creations stood outside his room near the cages, half completed and abandoned. "I started in the center making overlapping scales, then I realized I was going the wrong way, not enough wood left to make the arms," the man explained. This alien experience was short-lived; all three of us were fired when a wave of distemper wiped out a good part of the cat colony. A threatening letter from a client expressed the intention to "get rid of the rocker, the witch, and the whittler" and the displeased patrons made good on their threat.

My telling of this epic tale drew roars of laughter from Vaughn and his graduate students. "Why didn't you tell me you needed some work?" Vaughn asked. In truth I don't know why I hadn't, unless I was a bit embarrassed to admit that I wasn't yet making piles of money in the music business. Vaughn made a connection to the Los Angeles County Natural History Museum, where his respected recommendation actually got me a job. As though in compensation for suffering persecution from nuns eleven years before in elementary school, I was offered employment at the La Brea Tar Pits. I was hired as a pit boss to supervise the excavators on weekends and also to identify and catalogue—the title of curator is used for such scientists—the many bones extracted from the pit.

The project was actually managed by a sour old buzzard named George Miller, who had rubbed some of the museum administrators the wrong way with his bluntness and inflexibility. Miller had a kind of desert charisma, though; with the wrinkled brown skin of a chuckwalla and short gray whiskers, he stared at me with bulging eyes, looking for all the world like a half-starved, weather-beaten Humphrey Bogart in *Treasure of the Sierra Madre*. My meeting with Miller was a little tense; I could tell he felt pressured to take on a young academic type of little experience, and the interview was merely a formality he had to endure. In fact he treated me rather badly, grunting something incomprehensible about how it takes real elbow grease, not book learning, to be a paleontologist, and that young people today "don't know anything practical, because they're always watch'n' the boob tube." He tried to trick me by showing me a crazy-looking hand bone of a giant ground sloth, but was crestfallen when I actually identified the thing. I had anticipated this kind of test and scrutinized some of the La Brea bones on display at the L.A. County Museum beforehand. When the interview was over, he merely signaled the event with a wave of his hand, as if to sweep me out the door of his trailer into a tar pool. Nonetheless, as the weeks went on I gained a certain degree of gruff acceptance from George, and we became cordial, if not close. For many years afterward I would encounter him at the annual meeting of the Society of Vertebrate Paleontology. He always gave me a smile and would compliment me if I'd given a talk, as if he were proud of our earlier connection. I missed that grizzled old man at the meetings after he died.

By 1970 the tar pits had been excavated and studied for more than eight decades, and a lot then was known about the prehistoric life of La Brea. The most fascinating of all the creatures from this locality, which is now thought to represent an age span between 9,000 and 30,000 years, is a small woman. At first it seemed that the existence of human remains in this ancient fossil assemblage, reported by John C. Merriam in 1914, might be a fluke. But the bones of La Brea woman have been dated and redated by measuring the amount of radioactive carbon—carbon 14—in the bone.

Carbon 14, a form of carbon with fourteen neutrons in its nucleus, spontaneously disintegrates, or decays, at a constant rate but in living organisms is replenished at a constant rate that balances the loss. The new carbon 14 comes from the bombardment of cosmic rays from outer space, which alter the chemical elements in the body. When an organism dies, the radioactive carbon 14 in its tissues continues to disintegrate but now it is no longer replenished by the absorption of new carbon 14 from living tissue. After 5,600 years, a dead organism will have only half the amount of carbon 14 it had when it was alive—this is the half-life of carbon 14.

Since the loss of carbon 14 increases over time at a constant rate, the traces of the isotope in a given object can be used to pinpoint an ancient date with some accuracy. It is not good for very old bones or tree stumps because carbon 14 disintegrates too rapidly—after all, half of the carbon 14 is gone after only 5,600 years. So carbon 14 works best for dating the age of things on the measure of thousands, rather than millions, of years. The La Brea bones are in the perfect time slot for radiocarbon dating, and, like the other skeletons from the site, La Brea woman has been dated at at least 9,000 years before the present. The discovery that La Brea woman was so ancient was a sensation. Even though there are claims for more ancient humans in North America, she is still one of the oldest uncontested human fossils found on this continent.

Of course, this find has spawned many questions and speculations, most of them unanswerable. How did she get stuck in the pool? Was she chased into it by predators? Or simply careless on an evening walk? Was she an early American human sacrifice? She even inspired some ridiculous fantasies. One day, as I sat by my desk in the work shed writing catalogue entries for a series of camel foot bones, I was greeted by a strange-looking man with dark, deep-set eyes, a sharp goatee, and a silver necklace with a fake sapphire amulet. He claimed to be a movie director developing a project called "The Beast of the Tar Pits." I recommended saber-toothed cats, giant ground sloths, armored glyptodonts, and other fitting candidates for his title role, but he dismissed these. He had made up his mind, exclaiming, "There is no more horrible monster than a woman." As far as I know, "The Beast of the Tar Pits," with its terrifying female lead, is still a script in circulation.

La Brea woman is rather odd in a completely different way. She is the only evidence for humans in the Los Angeles Basin 9,000 years ago, while dozens of other mammalian species are represented by thousands of bones in this Tar Pit assemblage. One of the most common of these species is *Canis dirus*, the dire wolf. The scientific name here is in two parts: *Canis* is the genus, or group, to which this species belongs; *dirus* is the species name. *Canis* is a Greek word meaning dog, or cur. *Canis dirus* is thus a species with close relatives in the dog group, such as the wolf, *Canis lupus*, the coyote, *Canis latrans*, and the domestic dog, *Canis familiaris*.

There were big packs of dire wolves loitering around the pools, probably bringing down an old bison or scavenging the corpses of

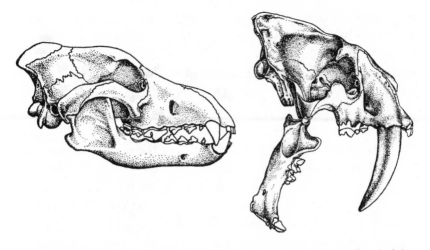

Canis dirus, the dire wolf (left), and *Smilodon californicus*, the saber-toothed cat (right)

drowned imperial mammoths. Indeed, hundreds of thousands of specimens, representing thousands of individuals, are listed in the catalogue in the Los Angeles County Natural History Museum. Accordingly, it is important to note that paleontologists don't need an entire skeleton to recognize an individual; one bone will often do. Some bones are better for this purpose than others. Two ribs, for instance, might come from two individuals or be different ribs of the same animal. True, there is—except in mythology—only one skull per animal. So why not simply count the number of skulls to find the number of different individuals in a pile of fossils? But skulls are made up of at least forty different bones (including right and left bones of the same pair), which are easily broken and dispersed when an animal dies and lies on the ground. First, the bones are picked over by scavengers and the skull is often broken apart in the process. Rains may then wash these skull shards downstream before they come to rest on a sandbar and are buried and preserved there. In fact, even isolated skull bones can be notoriously rare in a bone bed, and they probably grossly underrepresent the actual number of animals that once lived at the site. Fortunately, some other bones are unique to an individual and therefore highly useful for taking a census. Like dogs and cats, male dire wolves have a rod-like strut of bone in the penis, called a baculum. The large number of baculi catalogued in the L.A. County Museum collections attests to the abundance of male dire wolves around La Brea. Moreover, since all mammals show a sex ratio of roughly one to one—one male to one female in a population—multiply the number of baculi by two and you get a decent estimate of population size.

Identifying La Brea bones was interesting work, and I impressed some of the museum scientists who visited the site when I identified small nubbins of wrist bones down to the correct element and the correct species. The ground sloths were my favorite specimens because many of their bones had such bizarre shapes, and certainly those bones added up to an equally weird skeleton. It was also a thrill to meet scientists I had read about. One day Richard Leakey, the famous hominid hunter with the famous hominid hunter father, stood by my side nodding his head as I showed him different bones of juvenile and adult skulls of the La Brea lion, *Panthera atrox*, and the saber-toothed cat, *Smilodon californicus*.

I spent a lot of time counting bones and estimating the number of individuals that had been trapped in one particular ancient pool we had excavated. This procedure was actually well refined by that time. Between July 1913 and September 1915, the Los Angeles County Natural History Museum crews had made ninety-six numbered excavations. More than fifty of these were merely test holes and were completely unproductive. Examination of the state of the fossils and the nature of their burial allowed a recognition of the best sites. Only eighteen of the ninety-six excavations were self-contained concentrations, that is, discrete tar pools where big animals had fallen in and died. In seven of them the bones were in a poor state of preservation because they had been soaked in water. Of the remaining pits, two were used for display and never analyzed, one was very small with very few bones, and one contained only small species of which only the birds were studied. This left seven pits with rich, well-preserved concentrations of bone representing diverse animals.

Comparisons of the bones in these well-stocked pits revealed some variation in the animals that once lived there. In a 1960 paper, Leslie Marcus reported on a census of the best pits from the collection catalogues at the L.A. County Museum. Marcus used a standard technique for estimating the number of individuals, which was based on the maximum number of the same element in a given pit. The kind of element used for this purpose varied from pit to pit. For example, in Pit 3 there were 339 atlases of *Smilodon*, the saber-toothed cat—the atlas is the most anterior vertebra, which articulates with the back of the skull—and 245 right calcanea, a bone in the ankle (remember that in the case of these paired bones, one can only use a bone from either the right or left side, but not both, to estimate the total number of individuals). In Pit 4 *Smilodon* was represented by 91 atlases and 105 right calcanea. Thus, the maximum number of individuals for this species was taken as 339 (as indicated by atlases) in Pit 3 and 105 (as indicated by right calcanea) in Pit 4. Applying this method, Marcus found that plant eaters, except for the giant ground sloth *Nothrotherium*, did not vary significantly in numbers from pit to pit. On the other hand, the number of different carnivore species varied significantly. For example, the ratio of saber-toothed cats to dire wolves was 0.65 in Pit 3, but this ratio was 0.77 for Pits 61–67, 0.39 for Pit 4, 0.90 for Pit 77, 0.34 for Pit 13, 0.61 for Pit 60, and 0.33 for Pit 16. Why the differences? The pits were once pools that were probably not all continuously active at the same time. Different conditions may have pertained at different times—more *Smilodon* in the area tracking a herd of a particular species, or more droughts or perhaps even a random event such as a concentration of dire wolves at some other killing site.

What about the overall abundance of different La Brea mammals from all seven of the best pits? A selection of the most notable large mammals from the census Marcus took gives an informative tally:

Canis dirus (dire wolf)	1,646 individuals
Smilodon californicus (saber-toothed cat)	1,029
Bison antiquus (bison)	159
Equus occidentalis (horse)	130
Nothrotherium shastense (giant ground sloth)	76
Panthera atrox (La Brea lion)	76
Camelops hesternus (camel)	30
Total	3,146

As these results show, the dire wolf remains at the top of the abundance heap, with another flamboyant predator, the saber-toothed cat, coming in second. Also, the number of individuals of carnivores (*Canis, Panthera, Smilodon*) far exceeds the number of plant eaters or herbivores (*Nothrotherium, Camelops, Bison,* and *Equus*). This is surprising, because in a modern animal community big carnivores are at the top of the food chain and are far outnumbered by their prey, the herbivores. It takes a far larger investment of energy and production in the ecosystem to provide meat than to provide plants for food. When plant food is consumed and converted to muscle tissue in the bison, camel, or other prey animal, the converted nutrition and energy are in turn consumed by the top predators. In the case of La Brea, however, an explanation for the large number of top predators seems possible. The tar pits were a gathering area, like a small lake in the dry season in African Serengeti, for great masses of animals. Plant eaters could have fed too close to the pools and become mired—probably in most cases very young, old, or sick individuals. This was a prehistoric pigeon shoot for the carnivores: they simply hung around the pools waiting for the next herbivore to find itself awkwardly caught up in tar and pounced on the entrapped victim in a feeding frenzy. Since the target was so vulnerable, the predators rushed on the prey in crowds and of course also got stuck, in even greater numbers—either in the act of killing immobilized prey or as they scavenged on their tar-soaked corpses.

In our more modern excavation, we not only made estimates of numbers of individuals but also carefully measured the size of different bones to distinguish adults from juveniles or males from females. We also made exacting diagrams of the orientation of each bone, because such information is very useful for determining the direction of currents in the pools that may have carried the bones for some distance and scattered them. Concentrations of bones—for example, where the skull of a La Brea lion was mixed in with a large assortment of dire wolf skulls, sloth toe bones, and camel ribs—are called lag deposits, because the current that collected the bones kept running on while the bones themselves were entrapped by some obstacle, perhaps a rotting log or a depression in the bottom of a shallow pool. Finding one of these lag deposits was a great moment in the excavation of the pits. "I've hit bone—lots of it!" someone would shout, and we would gather around the discovery to give advice on excavating the pile of fossils and to celebrate.

This pit work required tolerance for filth, not the "clean filth" one finds when working sandy outcrops in an unpolluted wilderness, but a miner's trial in a dark oily pit under a smoggy sky. We became encrusted with a rind of oil, asphalt, and dirt, which even after a bath stained the pores of the skin. We were also exposed to trichlorol ethane, a potent chemical that we used to soak the bones to remove their tar coating. The noxious vapors from the soaking vat invaded the hairline tracts of my nostrils. This was of course long before the days of OSHA-approved procedures. But the work itself wasn't the pits. It brought daily excitement and satisfaction— a *Smilodon* canine tooth suddenly exposed like an icicle in the sandy tar, a pile of camel bones emerging in the corner of the excavation. It kept me thinking about the Los Angeles of thousands of years ago.

II

. . .

The Age of Mammals Revisited

I worked at the pits for several months, until graduation brought my stint at La Brea to a logical end. I was offered permanent employment at the place, but I couldn't see prolonging the experience, and without further training, I would not have much opportunity to move on. Paleontology, like other sciences, requires much learning and, particularly, original research beyond a bachelor's degree. If I was going to do this to my satisfaction, I might as well do it whole hog. Or should I? That dilemma tormented me. Was I really ready to give up my wild dreams of becoming the next big thing on the L.A. music scene? The band now had its sights on better jobs, in places where we might get exposure to the elite, where we could hang out and talk music with people like Neil Young or Steven Stills. One of our favorite artists of this sanctified community was the brilliant Graham Parsons, lead singer of the Flying Burrito Brothers, a band reconstituted from among relatively unknown musicians and established ones like Chris Hillman, former bass player of the Byrds. Parsons died a haunting and mysterious death one night in the Mojave Desert, and a musician who worked in our band replaced him in his high-profile band. It seemed we were circling ever closer to the anointed ones. We were part of that scene and if the possibility of recognition was still remote, it seemed at least plausible.

Then things started to unravel. Nick, the lead singer, wanted to extend his experience as a USC film student into a career as a Hollywood director. My brother Steven and I joined a more prosperous band, but it didn't jell; he seemed to fit in, I didn't. I was out on my own, doing odd jobs with one band or another. Our experience with a major record company was unpleasant and ultimately a failure. Many of the people we encountered in various situations were slime.

Vaughn was helpful and sympathetic. He suggested I immerse myself slowly, perhaps take a master's degree elsewhere, and return to UCLA nearly fledged and committed for a Ph.D. Despite a straight-A average in my senior year, it was clear I was not the most dazzling prospect for a graduate program at UCLA; mediocre grades in chemistry and German blighted my record. Vaughn mentioned that a talented young professor, Jason Lillegraven, had just been appointed at San Diego State University and was looking for a student. The school was not top-notch scholastically, but it offered well-rounded training in biology and geology. Besides, I thought, it was close enough to Los Angeles for those odd music jobs. San Diego was also easy to escape if I suddenly decided to pull up stakes and head north to the metropolis to resuscitate my music career.

This lukewarm acceptance of an offer from San Diego State ensured that my labors as an urban paleontologist would continue. Interesting 46-million-year-old mammal fossils had been found many decades ago in exposures flanking what the paleontologist who discovered them, Chester Stock, called "beautiful Mission Valley." This broad depression runs to the coastline just north of downtown San Diego. In the early 1900s, fossil-filled outcrops of the valley were subtly exposed in gullies filled with chaparral that harbored bobcats and orioles. Pleasant stands of trees fringed marshes populated by shore birds and elegant cranes. There was even a good fossil locality on the grounds of a beautiful old mission. By the 1970s, Mission Valley had become a giant axon of commerce for the rapidly expanding city. It had been thoroughly trashed, invaded by metastasizing shopping malls and freeway interchanges. But the freeways required big road cuts which, conveniently, exposed fresh rock of the very same layers that had produced fossils for Stock and his crews decades earlier.

The rocks themselves were important: they represented a decent slice of time called the later Eocene epoch, when a North America 46 million years ago was populated with very interesting experiments in the early evolution of horses, rhinos, tapirs, camel-like forms, primates, rodents, and even tiny, inconspicuous shrews and hedgehogs. The best-known Eocene localities were in the classic sequences exposed in or near the Colorado Plateau, places like the Uinta Basin in northern Utah. The coastal, later Eocene fossil beds of southern California were real outliers, and their fossils showed some differences from the animals that had lived in the interior. Much like Vaughn's query for Permian animal communities, Lillegraven wanted to know whether these coastal Eocene faunas were very different from those that thrived in the heart of the continent. To do this, Lillegraven planned to scour those Eocene road cuts for more and better bones.

Up to this point, my familiarity with fossil mammals extended only to the La Brea forms. These were notably big and sometimes rather weird, like the ground sloths, but all in all they were mammals not unlike those still alive today. Most of them were simply bigger versions of elephants, camels, and lions. As one might suspect, mammals become less familiar and in an evolutionary sense more interesting the farther back one goes in time. Mammals are indeed very old, as old as the dinosaurs. They first appeared in the Late Triassic about 200 million years ago in forms no bigger than field mice, and they showed interesting differences in the construction of the jaw and the back of the skull that separated them slightly from their synapsid relatives. One unique feature of mammals is the presence of three tiny bones in the ear region, the malleus, incus, and stapes, or the hammer, anvil, and stirrup, which help to transmit sound vibrations from the eardrum to the inner ear. In synapsids, however, the malleus and incus were actually small bones in the back of the skull and the lower jaw. The transformation of the synapsid jaw and skull bones (the articular and quadrate) into mammalian ear bones is one of the great scientific case studies of paleontology.

Another aspect of the skeleton that distinguishes mammals is their teeth. In reptiles like lizards and crocodiles, the many teeth along the line of the upper and lower jaws look very much alike. They may be sharp and knife-edged, cone-shaped, or peglike. But in mammals there is clearly a difference among teeth according to their position in the jaw and their use in obtaining and eating food. At the front of the mouth, incisors are used for nipping at prey or pulling out plant parts. The incisors may be modified into long, ever-growing, chisel-like structures, as is the case in mice, rats, and other rodents. This kind of incisor is ideal for breaking open seed pods, cracking nuts, or slicing through tough plant stems. Behind the incisors are the canines, very sharp and enlarged in meat-eating forms like lions and dogs. The obvious use of canines for stabbing prey and slicing up meat is enhanced by their position in the jaw where a very large bite force can be exerted. The force is great because of the lever arm distance of the canine tooth from the back of the jaw, at the jaw joint. (The lever arm advantage is the mechanical advantage one relies on when using a pry bar to overturn a log or a tire iron to unscrew the lug nuts of a car wheel.) In some plant-eating animals, like camel, deer, and cows, canines are not of much use and in these forms they are small or absent altogether. Behind the canines are the premolars, which further serve to break down the food. At the back of the mouth are the molars, which usually have large crown surfaces made up of cusps and troughs that fit, or occlude, against the opposing tooth that it meets when chewing. The molars are ideal for mashing (the more technical term is "macerating") food into a mush—a soft, saliva-soaked pulp that can be easily swallowed. Some mammals, like modern horses, have very specialized flat-topped molars that keep growing throughout life, much like the incisors of rodents. As the horse

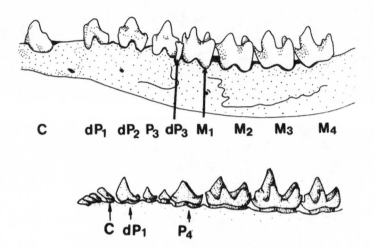

C   dP₁ dP₂ P₃ dP₃ M₁   M₂   M₃   M₄

C dP₁   P₄

Lower jaws of marsupials (above) and placentals (below), showing the differences in the number and replacement of molars (M), deciduous premolars (dP), and permanent premolars (P). The canine tooth (C) is also labeled

molar is worn away from chewing tough, silica-infused grasses, the molar keeps growing up out of the jaw to maintain an effective grinding surface.

This more complex battery row of teeth in mammals is different in another important way from those of reptiles and synapsids. In the latter, teeth are frequently replaced; many generations of new teeth pop out of the jaws as old ones are shed. Some forms may have several generations of these replacement teeth. In contrast, mammals have only two dental generations, the deciduous (referring to their tendency to drop out early in life), or milk, teeth and the permanent teeth. Moreover, not all sectors of the tooth row are subject to replacement. Incisors, canines, and premolars are usually deciduous and are eventually replaced by permanent teeth at the corresponding position in the jaw. In placental mammals, the very diverse group that includes rodents, lions, and our own species as well as other primates, all premolars that first erupt in the young are usually replaced by permanent teeth. In opossums, kangaroos, koalas, and other marsupials, only the premolar at the third position is replaced. Molars have only a single generation and usually erupt relatively late during development. Marsupials customarily have

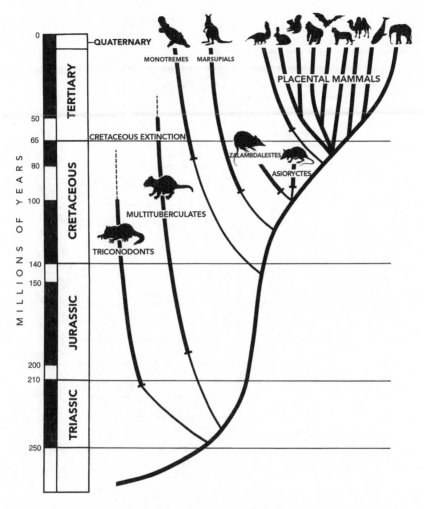

Some major branches of Mesozoic and Cenozoic mammals

four molars on either side of a jaw instead of the usual three in placental mammals. This difference between marsupials and placentals
with respect to the teeth—what is known technically as the dental
formula—as well as the mode of tooth replacement was recognized
by zoologists more than two centuries ago.

The earliest mammals with their three ear bones and differenti-
ated teeth lived literally in the shadow of the dinosaurs. During the
three chapters of dinosaur domination, the Triassic, Jurassic, and
Cretaceous Periods, the mammals remained small ratlike, shrewlike,

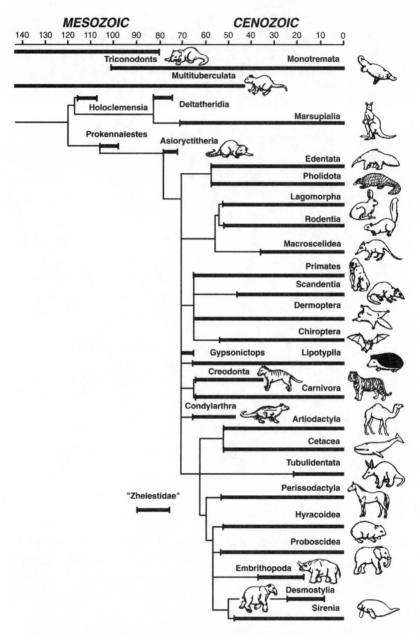

The placental mammal radiation in the Cenozoic, showing geologic age ranges indicated by fossils (thick horizontal lines) and branching events (thin lines)

or opossumlike forms, none larger than a gopher. We cannot ascertain much about their behavior, but it's likely that many of these early forms were nocturnal, to protect them from predators, like many mammals today. At the end of the Cretaceous, about 65 million years ago, the earth was shaken by a catastrophe that wiped out all the dinosaurs (save the birds) and many other forms. Many believe this worldwide extinction was caused by a massive asteroid impact near what is now the east coast of Mexico, although the exact pattern of extinction and its cause are still matters of much research.

The period following the Cretaceous and its extinction event is the Tertiary, when mammals finally got big and diversified into the many dramatic forms—from aardvarks to zebras—we see today. The Tertiary is further broken down into epochs because the mammals and other fossils known from this time interval so definitively record changes in the biota. The Tertiary epochs, running from oldest to youngest, are the Paleocene, Eocene, Oligocene, Miocene, and Pliocene.

The first of these Tertiary epochs, the Paleocene, was one of gradual, followed by more dramatic, change. In the early Paleocene, mammals remained small and much like their relatives in the later Cretaceous. For a surprisingly long time, as much as ten million years, there were virtually no medium-sized or large land animals browsing the leaves of trees and high bushes, an adaptation so well elaborated by the massive, plant-eating dinosaurs of the Jurassic and Cretaceous. At the same time, the Paleocene saw the radiation of many small mammals that specialized in fruit and other diets. These were the source of tree-dwelling groups like the primates. The radiation of these forms is important ecologically, because fruit-eating (frugivorous) mammals and birds are a primary mechanism for dispersing seeds of fruit-bearing plants in the spread of their own feces. Thus the radiation of mammals had corresponding effects on plant life. Both the mammals and the plants "took flower." Toward the end of the Paleocene, a few larger forms emerged. These included some piglike animals with chisel-like teeth, pantodonts, and some primitive meat-eating carnivorans, such as miacids, which looked like long, slender dogs with short legs.

The Paleocene also experienced some global warming, which was further accentuated in the first phase of the succeeding epoch, the Eocene. In fact, the early Eocene was a time when the earth had perhaps the highest global average annual temperatures for the last 150 million years. Mammals from New Mexico looked very much like mammals from the Arctic islands off Greenland. There is also other evidence for surprisingly warm and humid climates in the polar regions—turtles, crocodiles, and broad-leafed trees. During the early Eocene, about 55 million years ago, we see the first evidence of many modern groups: horses, tapirs, the dog and bear families, ancient toothed whales, bats, and the cloven-hoofed artiodactyls which later diversified into deer, antelope, camels, and many other groups. The early Eocene was then for some reason a great staging phase for the evolutionary radiation of these modern forms. In the later Eocene, an important climatic cooling and drying occurred and mammals and their habitats also showed a marked transformation. Habitats became more open and forms became larger, varying from browsers to types adapted for grazing in the spreading grasslands.

This trend continued through the Oligocene and Miocene, when great grassland and savanna habitats fostered herds of long-necked camels, horses, and thunderous rhinos and rhinolike brontotheres. The late Eocene was therefore an important evolutionary and ecological time zone, a time that saw the transformation from more closed to more open habitats and a concomitantly dramatic change in the mammals that lived in those habitats. Most of this transformation was recorded in rock sequences in the Rocky Mountain states and in western Europe and Asia. It was therefore also important to examine the nature of the change in other places, like the western coast of the North American continent, which had until this point only provided spotty evidence of this history. San Diego was an unexpected opportunity, a cityscape that held within it an important secret from the age of mammals.

12

Road Cuts and Fossil Vermin

B ecause of its relevance to studying mammalian history during late Eocene time, San Diego with its sporadic outcrops of fossil beds was an enticing target for field exploration. There was, however, a twofold challenge in fossil hunting in this urban realm: of course one had to find fossils, but first one had to find enough exposed rock to look for fossils. The lack of good outcrops drove us to desperate acts. We would see a tempting exposure of white Eocene rock sloping down into a set of suburban backyards and invade like a platoon of street urchins, only to be chased off by an extremely hostile lady of the house. We got clearance to quarry a road cut on the Camp Pendleton Marine Base. This locality—within a stone's throw of Richard Nixon's West Coast "White House" in San Clemente—proved quite rich but presented less than ideal conditions for excavation. One hot afternoon we were engulfed in the overpowering, oily fumes of burning creosote and manzanita.

We saw black smoke billowing over a hill above our quarry. I ran up to take a look and saw a patrol of war-painted marines charging us with flamethrowers and automatic rifles. It seems that the base officers had overlooked our existence when they moved their war games. Fortunately, the marines banked slightly and drifted farther north, scorching the earth as they went.

So we worked in this rather uncongenial locale, pulling ice plant away from a steep hillside to get a better look at the rock surface, plugging our ears and our nostrils when the freeway noise and the car fumes became intolerable. Even the work itself had a mechanistic, rather artificial feel to it. When a few flecks of bone or a tiny tooth no bigger than a pinhead was spied in a small chunk of rock, we immediately began breaking into the soft weathered surface of the sandstone with pickaxes. We shoveled the loose rock into burlap sacks and hauled it back to campus where we let it dry for a few days in the predictably constant San Diego sun. We then soaked the dry rock in kerosene to break it down and put the treated product into makeshift wooden boxes floored by a fine-mesh screen. The next step was laundry duty. We shook those boxes in long troughs of water, to allow the sand grains and clay to pass through the sieve. If the original rock sample was productive, the remaining residue in the box contained fossil bones and teeth in addition to the larger pebbles that could not be filtered out.

Then the prolonged sorting phase began. A couple of tablespoons of the concentrate would be put on a small paper or plastic tray and sorted under a microscope at six-power magnification. The degree of tedium here was of course directly dependent on the richness of the concentrate. Some samples were surprisingly generous, with small intricate lizard vertebrae, shrew-sized jaws, and sharp, shiny teeth of carnivores sprinkled among the concentrate. Unfortunately, screen-washed concentrate was usually not so entertaining. A ton of rock could be dried, kerosene-treated, broken down, shaken and sieved, only to yield a half-dozen teeth and bone fragments.

Why this painstaking procedure? The screen-washing technique came later in the history of paleontology. It was developed only in the 1950s, as an efficient way of extracting important albeit rather

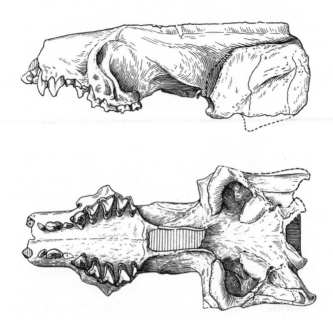

A side view (above) and bottom view (below) of the North American
fossil insectivoran *Apternodus*

inconspicuous fossils. It certainly worked in San Diego. We sorted
out thousands of specimens, snake and fish vertebrae, tiny limb
bones of lizards and small mammals, and shiny teeth of primates,
rodents, and, notably for me, insectivorans. Insectivorans today
comprise a number of small, sharp-toothed creatures—the hedgehogs of the Old World, shrews, moles, and a number of strange,
sharp-snouted, spiny animals called tenrecs from Madagascar. But in
the past insectivorans were much more varied and diverse: hedge-
hoglike forms and even tenreclike species flourished in North
America. These are admittedly not the most dramatic and certainly
not the biggest of ancient animals—they might be more aptly
called vermin than beasts. But insectivorans are very important in
an evolutionary sense. It was generally thought that they were the
group whose various lineages gave rise to all the orders of placental
mammals, the modern radiation that includes humans and other
primates as well as elephants, armadillos, bats, rodents, whales, carnivores, horses, and sea cows. It seems absurd to think of a whale as
transformed from a shrewlike form no bigger than the bristle end

of a toothbrush, but the fossil record was good enough to reveal a series of intermediate species that made this evolutionary history plausible.

Scientists were a bit intimidated by insectivorans, however. They turned up in appreciable and bewildering numbers, especially in sorted sediment, and many of them were unnamed and undescribed. Indeed, their taxonomy was a mess. As we continued sorting through our San Diego concentrate, it became abundantly clear that these enigmatic vermin were particularly populous 46 million years ago in the coastal scrubland of southern California. I suggested that I try to sort them out for a thesis, and Lillegraven was very encouraging. "I'll work on something easier, like the rodents, primates, and carnivores," he said.

My intensive focus on the diminutive creatures was not anticipated. In my early days as a fledgling paleontologist, I expected to continue working on big beasts like imperial mammoths and ground sloths, maybe even start describing dinosaurs, or at least scan skeletons as large as the eleven-foot *Dimetrodon*. But both Vaughn and Lillegraven made it clear that work on these big beasts wasn't necessarily an index of accomplishment. Many of the larger fossils had been described, and the small creatures often offered more evolutionary intrigue. They were the new frontier of research, items that provided remarkable insights on how bony structures evolved or how ancient animal communities were organized. In fact, both these early mentors shared a widespread disdain for much of the research on dinosaurs or big fossil mammals, which they characterized as superficial, driven more by a desire to make a splash on the public scene than by a wish to solve evolutionary questions. No doubt these attitudes made an impression on me; as I became more informed and more experienced in the field, I sadly encountered nothing that led me to disagree with this harsh judgment about such work. Only in more recent years has dinosaurology, for example, become more refined in a way that offers new, exciting, and reliable results.

Whether big or small, fossils weren't an easy target for analysis. I was a respectable collector, and I could write acceptable term papers reviewing an area or a problem. But now it was time to take hold of a specimen, put it under a microscope and brilliant fiberoptic light, and actually transfer what I saw there into words on paper. In looking back on that trial, I recognize the arrogance of a graduate student who thought he could easily dispatch these descriptions of cusps on teeth, of the line of the jaw, of the holes and troughs in a bone for blood vessels and nerves. Jay Lillegraven asked for a one- or two-page description of a single tiny tooth of a small shrewlike insectivoran, a posterior molar no bigger than a grain of rice. I wrote tentatively, rubbed out paragraphs, started anew, and labored to achieve the required precision. Is this cusp really taller than another cusp on the crown of the tooth? And if so, how much taller and from what view? Is the profile of the cusp cone-shaped or somewhat more swollen, and if so, how much? And, of course, were these features more or less expressed in this particular molar than in the same molar from another species? After an embarrassing two days I mustered what I thought was a presentable description. It came back bloodied with red ink.

The severe editorial hacking from Lillegraven was so devastating that I could not help rationalizing that he had intended it as a rude baptism, contrived to show me how much work lay ahead. But as I gradually improved in honing these descriptions of teeth and bones, I realized that my first try was indeed really lousy. Vaughn, quoting the well-known Harvard naturalist Louis Agassiz on the meaning of scientific research, used to say repeatedly, "Study nature, not books." This was of course only half true; even the most astute scientific observer could not ignore the relevant published scholarship. But, in the end, a great leap forward in science required not only a keen sniff of nature but a capacity for describing its subtle aroma. Sometimes a single flash, a vision of nature—a look through a pair of binoculars that convinced one that an unvisited line of red cliffs had the right sculpting and texture for fossil bone, or a perception that the eastern coast of South America and the western coast of Africa were so complementary that they could be fitted together like two snug pieces on a spherical jigsaw puzzle—could transform a career or a discipline.

I labored over my dissecting microscope, my eyes pressed so hard against the rims of the eyepieces that premature furrows were cre-

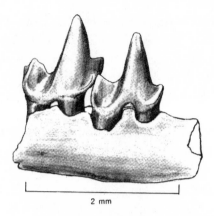

The first lower molar (left) and last lower premolar (right)
of the tiny Eocene shrewlike form *Batonoides*

ated. Under the magnified view between six and twenty power,
every crown of a tooth became a terrain. I surveyed each peaklike
cusp and followed its slope into valleys. These low elevations were
actually basins in the enamel designed to fit with the cusp of the
tooth above when a tiny beast some 46 million years ago moved its
upper and lower teeth together to pierce through a beetle shell.
When Lillegraven was not in the laboratory I would move to his
Zeiss microscope, whose finely grounded lenses and superb fiberoptic light etched the tiny world below as clearly as the Palomar
telescope would focus on the moon. Descriptions of myriad teeth
piled up over the months until I had a fat manuscript loaded with
tongue-twisting terminology and a batch of new names for previously unknown animals.

In making sense of all these minutiae I encountered another
challenge, one that comes from studying books in addition to
studying nature. It was not sufficient to describe the landmarks of
specimens and then simply compare them with similar-looking
teeth or jaws in other fossils. One had to demonstrate how these
features could help one draw a map of relationships among animal
species. The map was really in the form of a tree whose branches
linked one species with another, very like a family tree diagram. Of
course I was dealing with family genealogies on a grand scale. My
branches might link species separated by millions of years or con-

nect even bigger groups containing many species. This was not easy work because the characters had to be sorted out in a way that was consistent with the branching sequence.

Unfortunately, the published literature on fossil mammals was not very helpful in this tree-building procedure. Some papers did show in a rather precise way why certain features of a tooth or a jaw linked one species with another. But most were frustratingly vague and dogmatic. Too often, the analysis simply ended with a statement like "Clearly, such and such species has an ancestral-descendant relationship with such and such species." I couldn't figure why such a relationship was clear at all, and what evidence could actually be called on to make this statement. Clearly, indeed. One might as well survey a line of people outside a movie theater and claim that someone at third place in line was the sister of someone standing next to her, without noting the common physical features or mannerisms that might suggest affinity. The people who had written these papers seemed to presume that one should simply believe them without question, perhaps because they were recognized as experts on a group of insectivorans or early carnivores. Even the most influential paleontologist for many decades, George Gaylord Simpson, claimed that the practice of naming and connecting species and grouping them, a practice called taxonomy or systematics, was a matter of art rather than pure science. And he was certainly one of the culprits: he made frequent proclamations about the evolution and relationships of mammals that required pure faith. The more I read, the more this fiat and obfuscation drove me crazy.

Like a few practicing paleontologists and taxonomists of the day, I resolved to make things a bit clearer. I laid out the characters with detailed statements as to why a given species was more closely related to one than another. As I reflect now on the subsequent refinements in systematics, this first effort seems a rather feeble step. But I am grateful for the experience of trying to remedy the vexation and confusion that a legacy of literature on fossil insectivorans had created.

I was completely buried in the work of field collecting, curating specimens, studying, and writing in combination with teaching

anatomy labs and taking other classes. My absorption in these pursuits left little time or interest for other matters. My hair grew
down my back, and the right temple of my starkly black Buddy
Holly glasses broke off, framing my barely exposed face in an eerie
asymmetry. Students in one of my labs affectionately, albeit rather
derisively, nicknamed me Igor, but I was unruffled. For the first
time in my life I felt intensely involved in a creative, intellectual
journey, an exploration of a tiny sector of the fossil record and the
tree of life where I had left others behind.

My master's thesis was completed in the required two years. It
was big and esoteric; my dad joked that it could use a cover spicier
than the plain red leather binding embossed with gold letters. But
there were a dozen or so people in the world whose specialized
expertise enabled them to appreciate the work. Dr. Percy Butler, a
smallish English gentleman in his seventies with a brush mustache
above a perpetual pipe, who looked a bit like a character from *The
Wind in the Willows*, was in reality rather formidable, perhaps the
world's authority on the fossil insectivorans and related creatures.
During a museum study tour of the United States, he came
through San Diego, scanned my specimens, and carefully read my
manuscript. I was terrified of his reaction, but to my shock he was
very complimentary. Jay Lillegraven told me that Butler, in private
conversation, had said I had written a very impressive doctoral dissertation and was amazed to learn that it was only a master's thesis.
It was a moment to savor.

This positive reaction came with strong encouragement to enter
a doctoral program jointly sponsored by the American Museum of
Natural History and Columbia University. I also received overtures
from UC Berkeley, and Vaughn reminded me that he hoped that I
would return to UCLA. The Columbia acceptance was extremely
important because that program was perhaps the most famous one
in paleontology. It had, over a hundred years, fostered the work of
some of the most influential scientists in the field, and it could
offer the most broadly comprehensive collections of reptiles, dinosaurs, and fossil mammals of any museum. The Berkeley offer
proved very tempting, though: it fostered a great tradition in field
research and an impressive cohort of graduate students. And in ad

dition—not a trivial consideration—the paleo grad student offices in the Berkeley Geological Sciences Building opened to an enchanted forest of coastal redwoods, spruces, cedars, and cypress that draped over a panorama of the San Francisco Bay, the Golden Gate Bridge, and a fog-enshrouded Mount Tamalpais.

The Berkeley ambience was alluring. I tried to fight off my provincialism about California, but I remembered the bleakness and grime of New York City. I could already tell tales of woe and adventure concerning my first visit to Manhattan. After a late night of study in the American Museum's collections, I had taken the wrong subway and ended up at 1:00 a.m. in the South Bronx. Realizing my disorientation, I got out at a dingy subway stop and made a U-turn over the staircase to take the southbound D train back to Manhattan. There was no one on the platform save a stooped, elderly lady with a crooked cane, and, far behind her, a very large man. The man started walking toward me, and I realized he was completely naked except for an impressive Indian headdress and a vest of chain mail that appeared to have been ingeniously constructed from the rings of beer-can tops. He started warwhooping, but then I heard a rumble of salvation. The D train rolled up, and I got on. As the train pulled away I could see this urban chieftain turn and run, just ahead of two rotund transit cops, waving their nightsticks and laughing so hard they could barely apprehend their suspect. In reflecting on these and other moments I decided that I liked to work in New York but not live in it. I wrote Vaughn and expressed my intention to go to Berkeley to work on fossil vertebrates from the Cretaceous or Tertiary periods, at least 80 million years after, in Vaughn's own view, "everything interesting in vertebrate evolution had happened."

Once I made that decision, there were discussions as to whether I would continue my dedicated labors in the road cuts of San Diego. After all, there was much yet to find in urban Eocene rocks, as major achievements in subsequent years were to prove. In addition to La Brea, some of the greatest fossil localities in the world lie well within city limits. In Europe, a garbage pit near the German town of Messel produced a fauna of vertebrates a few million years older, but much more spectacular, than the San Diego Eocene. The

50-million-year-old Messel creatures are stunning in their com-
pleteness and filigreed detail of preservation. Shrewlike forms still
show the carbonized impressions of their fur. Bats are preserved
with butterflies in their stomachs, literally, a mash of delicate wing
impressions inside the abdominal cavity that can be seen with astounding sharpness in the image of a scanning electron microscope.
In another hemisphere, an extensive quarry of thin-layered shale in
northern China has in recent years produced the first stunning fossils of feathered dinosaurs. Despite these examples of glorious concentrations of fossils in convenient locations, I was ready to head
for the hills. I longed for those days in serene, empty Monument
Valley. Piles of bones and rich, screen-washed concentrate be damned.

Lest one finds this an undedicated allegiance to fun in the wild
instead of to scientific mission, I knew I could have it both ways. I
was filled with excitement by the opportunity for far-flung exploration. Berkeley was notable for some very active field programs,
including an exploration of the famous Cretaceous dinosaur and
mammal badlands of Hell Creek, Montana, and a long-ongoing
exploration in the barren deserts of Australia. There were people, even graduate students, working with Richard Leakey—
Leakey, no less!—in East Africa, and field parties striking out for
Patagonia. At Berkeley, students were encouraged, even expected, to
bring paleontology to the hinterlands.

The day I deposited my master's thesis at San Diego State University was a particularly clear one. The greenish brown smog that
usually hugged the inland freeways had been blown out by a dry
wind from the east. The Laguna Mountains, very like the San
Gabriel Mountains I savored from a distance as a boy, were sharp-
edged and deep blue. Where was the next fossil find of a lifetime?
What was entombed beyond the ranges?

13

. . .

Badlands, Bones, and Bone Hunters

In the summer of 1973, Professor Don Savage invited, nay expected, me to join his Berkeley team for an exploration of Montana and Wyoming even before I hit the books the succeeding fall semester. The prospect thrilled me. I hadn't been to the Rocky Mountains for three years—the last time with Vaughn in the Sangre de Cristos and in Monument Valley. Since then I had gained much familiarity with paleontology and even some certification, now that I had completed the master's thesis. It would be great to get back out to open country, this time not feeling like such a greenhorn. Part of the point of the upcoming season was to scout out various potential fossil localities. If one of these spots turned out to be tempting, it was understood that I could in subsequent seasons even take my own crew to work them. I stared at a geologic map of the northern Rocky Mountain states and drew imaginary circles around what might turn out to be my own fossil sites.

Before heading out, I first had to move the furniture my wife
and I had accumulated over two years and store it in a garage at my
brother Steven's house in L.A. One can easily resist recounting such
mundane activities, but regrettably the experience proved to be significant, not only robbing me of a field season but nearly killing me
as well. After a week of finishing up my San Diego obligations with
virtually no sleep, we drove a couple of loads of furniture to L.A.
during one of those outrageous 100-degree-plus heat waves that
come to southern California. As I was getting ready to return the
rented trailer to San Diego, a stranger with a blank though somewhat angelic face asked me for a lift. He seemed a little weird,
"spaced out," but I told him to hop into my father-in-law's
stretched-out Ford Country Squire station wagon.

The next sequence of events is a bit fuzzy. I remember passing
the rows of oleander bushes that line the San Diego Freeway as we
moved in tandem with the traffic coursing through Long Beach. I
was telling the stranger tales of Montana and about famous paleontologists who once roamed there in the late 1800s, when hostile
bands of Indians were still a danger. Then I noticed something odd
about Long Beach; it seemed to be floating above me rather than
alongside. My head felt very comfortable, as if it were nestled in a
soft down pillow, but I was beginning to lose contact with my ex-
tremities. Things went black. The Ford station wagon pulling an
empty trailer and going seventy miles an hour suddenly lurched
into the right lane. The stranger lunged forward in fear and snapped
me awake. As I guided the car to the shoulder of the road, I could
no longer feel the steering wheel. I could not move my arms or
legs or turn my head. The stranger was now a guardian angel; he
took the wheel and drove me to a nearby hospital, where I remember hearing a doctor say to someone, "Does this young man realize
he is in serious trouble?" I passed out for an unknown spell, then
suddenly awoke feeling agitated and very hot. A nurse said in a
tone of morbid detachment, "Blood pressure, two-ten over one-
twenty."

I did not seem to have any symptoms of an insidious health
problem; merely the arrogance of someone who believed in the
invincibility of his twenty-four-year-old body. The doctors judged

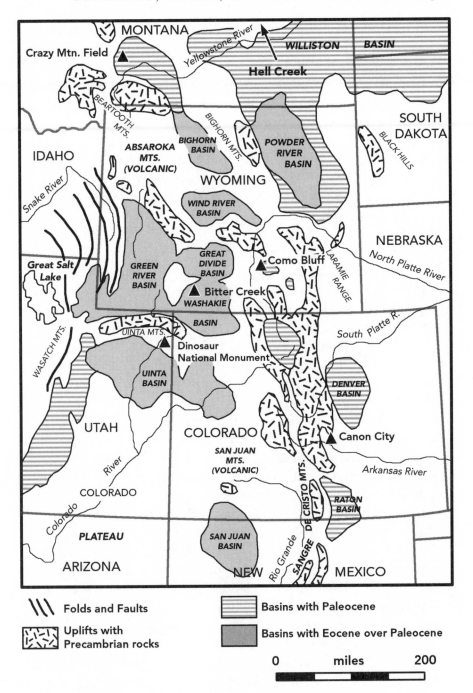

Basins and selected fossil localities in the Rocky Mountains

my condition to be critical heatstroke, a physical breakdown result-
ing from too much heat and too little sleep. Heatstroke has many
manifestations. The aftershock of this one had a nagging effect
of fraying the "wiring" of tiny nerves in the semicircular canals
around the inner ear. This is the area controlling the sense of balance, and I dealt for the length of a tortured summer with prolonged vertigo, a misery that kept me prone for weeks. It is hard to
fight off depression in these circumstances. I kept telling Savage I
would be all right soon and entreated him to leave a slot open for
me in his expedition. In early July, the day of departure for Savage
and company came. In a shaky voice over the phone, I told him,
No dice.

At the conclusion of this endless, bedridden summer, I felt strong
enough to make the move to the Bay Area. The year that followed
was difficult, as my dizziness made even walking the mile from our
apartment through the redwoods and cedars of Berkeley's north
campus to the geology building an epic trek. As the months passed,
however, the condition attenuated and I started to get my land legs
back. With delight, I joined Savage and his colleague, Professor
William ("Bill") Clemens, one year delayed, for another summer of
fieldwork in the American West.

Like the Colorado Plateau to the south, Wyoming and Montana
are full of mountains, deserts, canyons, and fossils. The hot spots for
paleontologists are located in a series of broad shallow basins or
badlands flanking river drainages. These eroded landscapes offer
cliffs of sandstone, shale, and limestone that represent a span of millions of years of time, extending from the early dinosaur reign to
the heyday of mammalian evolution. The rocks record the gradual
change in clime and topography that took place in the western part
of the continent. During the earliest phase of the age of the dinosaurs, the Triassic, much of Wyoming and Montana was covered
with a vast, shallow inland sea, which underwent periods of expan-
sion and recession. Many of the huge, thunderous, long-necked dinosaurs and their predators that lived some 150 million years ago in
the Jurassic Period roamed about large mudflats and shorelines that
bordered the seaway. These Jurassic dinosaur communities are well
preserved in the Morrison Formation, a thick band of rocks that

were once stream channels, floodplains, and lakesides and that are exposed through much of Colorado, Wyoming, and Utah. The fossil localities containing these monsters are rich and famous; they include Canon City, Colorado, Como Bluff in south-central Wyoming, and Dinosaur National Monument in Utah.

Some millions of years later, during the Cretaceous, a great northwest-trending seaway once again invaded much of western North America, cutting the continent into two parts. The dinosaur-dominated communities are especially well preserved on the western shore of this ancient sea. There, *Tyrannosaurus* once went fang and claw with the great horned dinosaur *Triceratops*. These near-shore Cretaceous communities were also populated with a great diversity of less imposing vertebrates—freshwater fishes, sharks, turtles, crocodiles, birds, frogs, salamanders, lizards, and mammals, which are represented by bones that poke out from the cliffs flanking the Red Deer River in Alberta. Farther south are more fossils in a maze of badlands that make up the Hell Creek Formation—especially well exposed south of the vast Fort Peck Reservoir in eastern Montana, which impounds the waters of the upper Missouri River. Cretaceous dinosaur beds pop up in Wyoming in many places, notably the buffy sandstones of the Lance Formation near Lance Creek, on the edge of the Great Plains in the eastern part of the state.

Somewhat younger Tertiary rocks with lots of fossil mammals are exposed in basins that fill great sweeps of land throughout western Wyoming and southwestern Montana. These rocks were laid down when the inland sea started to dry up in the early Paleocene, just after the great Cretaceous extinction of 65 million years ago. The layers of rocks in these basins, representing a complex recycling series of lake beds, floodplains, stream channels, and deltas, are so broad in extent that from their center the mountains at the fringe often look like no more than purple wisps of clouds on the horizon. Nevertheless, on a geologic map, the basins are clearly identifiable, like giant craters on a vegetated moon, and many of them can be located roughly in reference to the mountains that border them. The vast Green River Basin is flanked by the Thrust Belt Range on the west and the magnificent Wind River Range on the east; the Bighorn Basin by the Absaroka Mountains to the west, the Owl Creek Mountains to the south, and the Bighorn Mountains to the east. Likewise, other basins have other mountain ranges encircling them.

The geology and geometry of basins are well designed for prospecting. There is an orderly, elegant sequence of rock layers, and you can almost estimate the time zone the rocks represent by where you are in the basin. This is because the rock layers dip gently down toward the center of the basin and arc up toward the opposite side. Hence, the oldest rocks tend to be exposed on the outer fringes of the basin where the rock layers have been tilted sufficiently upward to reveal them. During a drive in the Red Desert down a potholed asphalt road south of Interstate 80 toward the ghost town of Bitter Creek, Wyoming, one first encounters the yellow-gray beds of the Fort Union Formation, which are late Paleocene—about 58 million years old. They tilt downward and in a southerly direction, and by the time you reach Bitter Creek they are overlain by a series of gray and blond sandstones and mudstones interbedded with small coal seams that contain small horses, primates, insectivores, rodents, and lots of turtles of the early Eocene about 55 million years ago. These are the lower units of the Wasatch Formation, which in some places are invaded by a shale layer of the Green River Formation. Continuing south down the Bitter Creek drainage, one catches sight of the next rock layer, the Cathedral Bluffs, a dazzling, candy-striped series of rocks that makes for a startling crown to the drab beds below. The beds get younger and younger until, in the center of the basin, you are surrounded by sandstone cliffs and spires studded with bones of big, rhinolike forms. These beds of the Uinta Formation represent the late Eocene, about 45 million years before present.

The basins and badlands of Montana and Wyoming are among the greatest fossil dinosaur- and mammal-bearing terrain on earth, as was soon realized, when the American West opened up in the late 1800s. Earlier, in the 1840s, naturalists had marveled at mysterious footprints in the Triassic slabs in the Connecticut River Valley of New England that eventually came to be identified as dinosaur tracks. In 1855 Dr. Ferdinand Vandiveer Hayden led an expedition to the Western territories and collected some fossil teeth near the confluence of the Judith River and the wide Missouri in Montana. When he returned east, Hayden gave these specimens to a prolific young researcher named Joeseph Leidy, who studied paleontology and anatomy at the Academy of Natural Sciences of Philadelphia. Leidy named these teeth as belonging to new taxa called *Palaeoscinus* and *Troodon*, and noted a resemblance with some "European reptiles" that were by then known as dinosaurs. This was the first formal recognition of dinosaur skeletal fossils in North America.

As Leidy continued to describe other dinosaurs, including the duck-billed *Hadrosaurus*, found in Haddonfield, New Jersey, these prizes became ever more alluring. The finds accumulated and the competition heated up. The era of bone hunting in the West that began in the 1870s was marked by a very bitter rivalry. One of the antagonists, Othniel Charles Marsh, had experienced a childhood in upstate New York of poverty and disoriented parenthood provided by his widowed father, a stepmother, and various uncles and aunts. The young Marsh enjoyed collecting fossils but made a lackluster showing as a student. He was fortunate to have an extremely rich and nurturing uncle, the famous philanthropist George Peabody. With the help and encouragement of Peabody, Marsh eventually matriculated at Yale University. His years there were marked by study trips to Europe, where he came to appreciate the natural history museums and their great collections of fossils and other objects. He came back with an ambitious and self-serving plan: he persuaded his uncle to give Yale a massive endowment for a new museum, the Peabody Museum, where Marsh himself would hold an endowed chair as professor of paleontology. Thus Marsh ended up as a rather aristocratic paleontologist in his own museum, unsalaried but generously sustained by a lifelong income from his patron uncle. At least in validation of his good fortune, Marsh did become a respected man of science; he published 270 articles and monographs in paleontology and filled his museum with massive, spectacular skeletons of dinosaurs.

Marsh's successful scientific career and insulated life was frustrated by the appearance on the scene of another ambitious paleontologist. Edward Drinker Cope, like Marsh, benefited at points in his life from family wealth, but in many other respects he was very different. Whereas Marsh gained in status and wealth, Cope lost much of his father's fortune in speculative investment and ended up depending on a modest salary from the University of Pennsylvania. Whereas Marsh was rather indifferent and plodding as a young naturalist, Cope was amazingly precocious; at the age of six he made sophisticated notes and reports about the ichthyosaur, an ancient sea reptile skeleton mounted in the Philadelphia Academy of Natural Sciences. These early passions were the roots of an extraordinarily prolific career. As a professor at Haverford College, Cope spent much of his life using the Philadelphia Academy as his home base, and produced a staggering 1,400 publications on his own rather eccentric version of evolutionary theory, as well as descriptions of recent fishes and reptiles, and diverse fossil vertebrates, including of course dinosaurs. Both Cope's productivity and his mastery of wide-ranging subjects in natural sciences far outstripped Marsh's. This divergence in monetary fortune and in scope of interests was to have a strong bearing on the outcome of the great game that Marsh and Cope played against each other over the vast dinosaur-bearing territories of the West.

At first, Cope and Marsh had a cordial relationship; in 1868 they spent a friendly week together poking around Leidy's old Haddon-

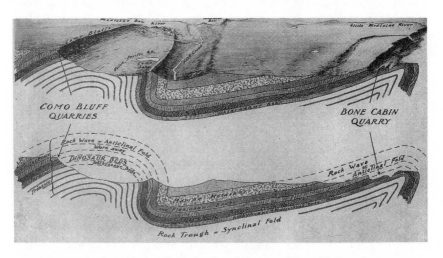

An archival geologic sketch and diagram of Como Bluff

field quarry for hadrosaur bones. The trouble started in 1877. Arthur Lakes, a schoolmaster and collector living in Colorado, shipped to both Marsh and Cope some huge dinosaur bones that he had found in the hogback foothills of the Rockies near Morrison, Colorado. Marsh sent Lakes $100 for the bones and told him to keep mum, but Cope, in keeping with his speedy publication pace, was already describing in scientific journals the dinosaur sample Lakes had sent him. Marsh then compelled Lakes to ask Cope to transfer the bones to Marsh, an outcome that infuriated Cope.

Revenge came in the form of another discovery of big dinosaurs from the same rock sequence near Canon City, Colorado. Much to Marsh's consternation, these even bigger bones were sent to Cope. Thus began a pitched battle over specimens and rushed efforts to be the first to publish the name of new dinosaurs. The confrontation, laced with "bone envy" and open hostility, lasted for more than a decade.

Later in the summer of 1877, the battle shifted northward when two collectors sent Marsh bones from Como Bluff, near the Union Pacific Railway line in south-central Wyoming. This site had a stupefying wealth of dinosaur skeletons piled on top of one another. (Many years later, fossils were still so abundant that a sheepherder built a cabin entirely from a scrap heap of dinosaur bones!) The crews hired by Marsh mined Como Bluff over the rest of the summer and even through the harsh and bitter winter of 1877–78, forcing the men to extract sauropod limb bones from snowdrifts in subzero temperatures.

The year 1877 was also a milestone in methods of fossil excavation. Samuel Wendall Williston, a medical doctor, was moonlighting as a pit boss in Marsh's quarry at Canon City. He recommended stabilizing dinosaur bones with strong paper soaked with flour paste, much as one might set a cast for a broken leg. The workers at Como Bluff employed plaster for this purpose, and thus originated an effective technique using burlap-soaked plaster for the protection of fossil skeletons—the same technique that we employed at Vaughn's Interval 300 Quarry and that virtually all bone diggers still use today.

The multiseasonal quarrying at Como Bluff continued over a

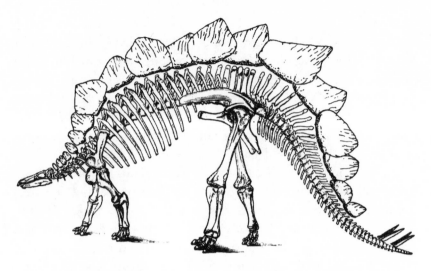

Stegosaurus, from the Jurassic of western North America, in a 1956 drawing

decade into the late 1880s. This bonanza attracted Cope's field parties in 1879 and 1880, an intrusion that Marsh clearly did not welcome. Legend has it that the Cope and Marsh teams had a pitched battle with fists and rocks at the site, but it seems likely that the extreme mutual hostility between the two paleontologists was not shared by their respective field teams. Marsh's men did actually propose leasing the land around the quarry to keep Cope's crew out. Meanwhile Marsh decided to invade Cope's digs at Canon City, Colorado, although it was apparent that the latter got the better of the big bones in that spot. Cope was at the same time distracted by other discoveries, including some at the rich Cretaceous dinosaur beds near the Judith River in Montana. Gradually the feud began to peter out. Unlike Marsh, Cope lacked the private funds required to sustain these massive excavations, and with his remarkably broad interests he was expanding his research on many fronts in zoology and paleontology.

In contests like these, the identity of the real winner is moot and perhaps irrelevant. Marsh ended up with more bones, but Cope was forging a monumental body of scientific work, of which dinosaurs were but a single aspect. What is relevant is that this frenzy of competition produced some of the greatest dinosaur skeletons

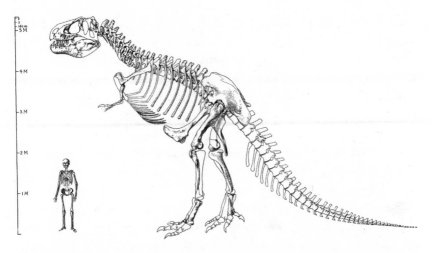

An old-fashioned reconstruction (from 1905) of the skeleton of the type specimen of *Tyrannosaurus*. Modern reconstructions show the tail rigid and high off the ground, the head and neck bent forward

ever recovered. Famous species like meat-eating *Allosaurus*, the long-necked *Brontosaurus* (in recent years renamed *Apatosaurus*), the plate-backed *Stegosaurus*, the single-horned, shield-headed *Monoclonius*, and many others all emerged from the furious competition between Cope and Marsh. And this race inspired a tidal invasion of bone diggers in the American West, most of them supported by the great museums of the East. Train cars full of huge, stony black bones, some of them resembling the trunks of incinerated oak trees, came trundling back to Chicago, Pittsburgh, Philadelphia, New York, and other cities. Awesome stacks of bones were laid out on museum floors, reconstructed, and built like giant Erector sets into dinosaur mounts.

Fortunately the dinosaur fossils of the Rocky Mountain states continued to be bounteous. In 1898, field parties from the American Museum of Natural History located the legendary sheepherder's cabin made of dinosaur bones near Como Bluff and promptly opened the famous Bone Cabin Quarry. By the end of one summer at this site, crews loaded 60,000 pounds of bones, which filled two freight cars supplied by the New York financier J. Pierpont Morgan. Over the six-year period of 1898–1903, American Museum crews working at Como Bluff and along the Medicine Bow ridge amassed well over 150,000 pounds of dinosaur bones to yield the greatest dinosaur collection in any museum. Crews from the Carnegie Museum of Natural History in Pittsburgh opened a rich suite near Vernal, Utah, which still amazes throngs of visitors today as Dinosaur National Monument. At about the same time, the indefatigable dinosaur hunter Barnum Brown explored Alberta's Red Deer River and the Hell Creek Badlands of Montana, bringing to the American Museum's collections an unrivaled assemblage of Cretaceous dinosaurs, including the famous pair *Tyrannosaurus* and *Triceratops*.

This history of desperate searching, feuding, and triumph was the founding legacy for modern field paleontology in the great basins and badlands of Montana and Wyoming. By the 1970s these areas were more overrun than ever by professional paleontologists and amateur and commercial collectors. Though the open range wars of the Marsh and Cope days were over, a certain rivalry persisted, a "spraying of the stump." Berkeley crews worked at the northern edges of the Washakie Basin in the older stuff—the Paleocene and early Eocene—whereas the team from the Field Museum of Natural History in Chicago worked the center of the basin in the upper Eocene section. Berkeley crews also worked extensively in the Cretaceous and Paleocene badlands of Hell Creek and were moving in on a rich Paleocene site in southern Montana known as the Crazy Mountain Field, originally excavated in the

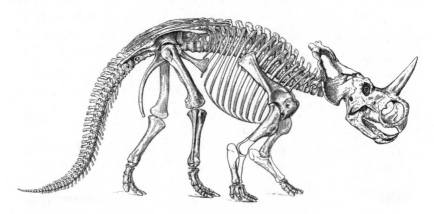

Monoclonius, as drawn in 1916

1930s by field crews from the American Museum. The latter institution now laid claim to the Cretaceous Lance Creek and some choice Eocene localities west of the Wind River Mountains. Carnegie Museum crews fanned out in the Green River Basin, sampling the Bridger Formation. The magnificent, seemingly endless sequence of badlands in the Bighorn Basin were largely the territory of a cluster of institutions linked by traditional academic ties involving professors and their students at Princeton, Yale, the University of Michigan, Johns Hopkins, and the University of Wyoming. The Hell Creek, the Bridger, the Washakie, the Lance, the Bighorns—these were still the fields of dreams for the new bone hunters, places that inspired planning, possession, and ultimately plunder. Modern bone hunters worked in the West with Marsh and Cope in mind—perhaps with a trifle less avarice, greed, and aggression, but with as much zeal and longing for the good stuff—all in the name of science.

14

. . .

Hell Creek Is for Dinosaurs

The Berkeley plan for my inaugural summer in Montana and Wyoming called for a reconnaissance of a few historic localities in Montana. We then planned to join Professor Clemens's party at Hell Creek and later head south for more scouting in Wyoming, until we ended up in Washakie Basin, in the far southwestern corner of that state. En route we headed over the cool Sierra Nevada and then downward and eastward into broiling Reno, Nevada. From there we took a jog north toward Winnemucca, Nevada, a forlorn town encircled by trailer camps implanted on a scarred desert plain. For camping that first night we decided to take to the spectral, gray Pyramid Mountains west of town. A couple of other grad students and I went for a short hike up the cool creek near camp. It was a pleasant walk, on which the Nevada heat was tempered by the 8,000-foot elevation. I had a powerful thirst and I bent down to scoop up the icy water, then immediately buried my face in it. As I slurped water like a desiccated camel, one of my mates laughed and kicked me at the shoulder. He pointed to an archipelago of sheep shit floating downstream. I promptly stopped drinking, but by that time I had already sucked up about half a gallon of the infected creek.

This was the early 1970s, when most backpackers still drank creek water with confidence and mountain wilderness trekking guides gave them only mild warnings. I hadn't even heard of gravity filters or other such paraphernalia for purifying water. After all, water from a mountain spring was so clear and cold, so good to the taste, it had to be safe. There was a lot of talk among hikers, however, that giardiasis and nasty bacterial diseases that caused even higher fever and more severe diarrhea were mysteriously spreading. Several days later I could be counted among the victims.

The first paleo stop on our tour was not a place for dinosaurs. In southwestern Montana there is a dark, pine-encrusted range that retains its snow even when the grasslands at its footings are singed by the July sun. This magnificent set of peaks is called the Crazy Mountains. Don Savage guided us over the pretty, rolling grasslands below the Crazies to gullies of sand, craters no bigger than a handball court (handball was Savage's favorite recreation). But these sites, located in a terrain aptly named Crazy Mountain Field, were actually famous quarries chock-full of mammal fossils that had been mined in the 1930s by teams under George Gaylord Simpson from the American Museum of Natural History. One of these bone pockets, called Scarritt Quarry, looked like an ancient pond deposit layered with pearly shells of freshwater snails and clams. Interbedded in this pearly layer were rare but fine jaws of mammals from the Paleocene, about 58 or 60 million years old.

Savage was a debonair fellow in his early sixties with a shock of silvery hair, a distinctive Texas drawl, and a formidable mastery of all things paleontological. I had first met him when he spent some time in San Diego as a visiting professor while I was a student, and we developed an easy, casual relationship. We hunkered down in Scarritt for a couple of days, shoveling off overburden and carefully turning over the layers of shelly shale and sandstone with the broad blade of a knife. As we quarried, Savage reminisced about his days

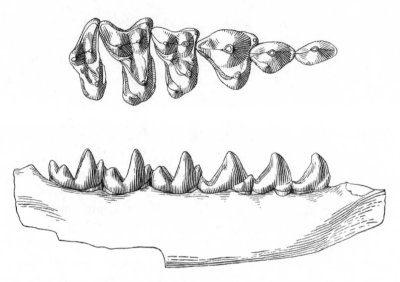

The upper (above) and lower (below) teeth of *Bessoecetor*, a small Paleocene mammal from Crazy Mountain Field, Montana

working with his former student Don Russell in the Paris Basin, where two centuries earlier Charles Lyell had made his famous calendar of geologic time based on the sequence of rocks and their fossils. He kept referring to the shell beds of Scarritt as a *coquillage*, and tortured us with other French terms spoken with a distinctive Texan twang. "The work, and certainly the wine, is more civilized in good ol' Franshaaaaze," he would say.

Crazy Mountain Field had its own pastoral charms. We stayed in a complex of ranch buildings snuggled in a small vale. In the evenings the pines on the low hills nearby cast long fingers of shadows in the dying light. The morning came with the slap of brisk air that seemed to sweep down from the Crazies, only to be beaten back by the burn of the morning sun. As usual, I liked to sleep outside, bedding down on a dry, warm evening and awakening in an ocean of dew. It took a couple of hours for the sun to dry out my sleeping bag. After suspending my bag on a corral railing, I would take a short walk along the range fence. At regular intervals, grasshoppers were impaled on the barbed wire wrapping the fence posts, the work of some shrike. Strolling back toward the ranch house I could hear the soft yawns and conversations of my field mates.

Among these was Howard Hutchinson, a reserved, bearded man in his late thirties. Howard was a consummate collector and an exacting paleontologist. He was not a professor at Berkeley, although he certainly had the expertise to qualify him for the job. Instead, he was a museum scientist in charge of taking care of Berkeley's superb vertebrate fossil collections of 135,000 specimens. He was not at all anxious about this status, or desirous of changing it. For an excruciating several years he had worked earnestly but glacially on a Ph.D. dissertation on the taxonomy of fossil shrews. In the meantime he fell victim to various distractions—including turtles. These lowly creatures left shells everywhere, through every fossil-bearing rock formation in the Cretaceous, Paleocene, and Eocene. But most of these shell pieces were nothing more than shards, from which it was impossible to recognize the species that once had the part in the mosaic of its shell. Undaunted, Howard mastered the rare skill of identifying not only the part of the shell from which a little fragment had come, but also the species or at least the genus to which the animal had belonged. In this way he had established a sort of turtle-shell timescale. He could often tell you were in the Late Cretaceous or the middle Paleocene by the kinds of turtle-shell bits in the rock you were standing on. This talent put Howard much in demand, and he was constantly being asked to join expeditions in faraway places, including the islands of the Arctic, to provide his expertise on turtles and their place in time.

Howard's popularity and his opportunities to travel taught a good lesson to those who dreamed of being chosen for international expeditions among dozens of fledglings claiming an interest in dinosaurs or early hominid fossils. Such expeditions need real experts, not just enthusiastic glory seekers. The romance and triumph would come in due course. Howard also had other consummate interests, especially sci-fi and horror books and movies. As we sat around the campfire under a sickly, yellow half-moon, he terrified us with a scene-by-scene retelling of *Night of the Living Dead*, one of the most classically gruesome movies of all time.

Pickings were slim at Crazy Mountain Field. We had a few nice jaws and skull parts of some ancient snail-eating mammals to show for a few days' effort. But it was clear that with enough digging and scraping here and at nearby Gidley Quarry one could probably come up with a decent sample of stuff. Savage kept saying, "This would be a good dissertation project for you, Mikey John," but I wasn't enthusiastic. How could a new project improve on the productive excavations that had gone on here nearly four decades ago? I felt like asking him, If it's so good why haven't *you* worked here yet? His urging seemed to run against the grain of another common paleontological adage he often uttered: "A locality is never again as good as the first time you find it."

Despite its green pastures and Swiss alpine vistas, I was glad when we deserted the Crazies after a few days and headed north and east to the Hell Creek Badlands. We were in real cattle country now. Here were huge tracts of grazing land and ranches as big as some states. Towns—if you could call them that—were merely supply depots that drew in ranchers and farmers from hundreds of miles all round. The air rushing through the window of our yellow UC Berkeley pickup truck was a mixture of sage, wild onion, clover, and manure. After a drive of endless monotony over a flat, fertilized earth, we stopped in Jordan, a junction only a few miles south of Clemens's base camp in the Hell Creek, and apparently one of the most isolated towns of any size in the contiguous United States.

The Hell Creek is famous not only for its dinosaurs. True enough, the type, or reference, specimen of *Tyrannosaurus rex*, the magnificent skeleton in the American Museum of Natural History, was discovered here by Barnum Brown, the swashbuckling paleontologist who wore a boot-length fur coat and carried a rifle. Brown and his crew found that incredible specimen in 1902, but there was still plenty left in the maze of rounded buttes and soft sand gullies that stretched to the horizon. The Hell Creek is also unique as a rock sequence. The rocks grade upward, through Cretaceous beds, known as the Hell Creek Formation, that preserve the last dinosaurs, into higher Paleocene rocks, known as the Tullock Formation, where dinosaurs are absent. In these upper rocks instead are small mammals—insectivorans, opossums, and a few small progenitors of herbivorous hoofed forms called condylarths. In this transition there is also a radical change in plant life, as indicated by the fossil pollen grains preserved in the rock.

The boundary between the Hell Creek and Tullock rock units is a thin black line, actually a seam of carbonized black coal, known as the Z coal. This is merely the lowest of the coal seams that typify the Paleocene Tullock Formation; such a complex of coals is sporadic or lacking in the Hell Creek Formation. The horizon for the extinction event within this sequence is, however, a bit fuzzy. Dinosaur fossils peter out a few feet below the Z coal and typical Paleocene mammal fossils and pollen changes appear a few feet above the Z coal. This so-called ten-foot gap in fossils has attracted much tortured thinking and debate. Despite the mysterious gap, the Hell Creek Formation with its dinosaurs, topped by the Z coal and the Paleocene Tullock Formation above, preserve a vivid transition from the age of dinosaurs to the age of mammals. This transition for fossil vertebrates is in fact more completely represented in Hell Creek than anywhere else in the world.

Clemens intended to chronicle the change in fossils—of dinosaurs, crocodiles, turtles, and mammals—in kind and in numbers through the Hell Creek–Tullock section. To do this he worked with a squadron of colleagues and grad students who specialized in geology or in different kinds of fossil. One of the most important events in earth history—the end of the dinosaur age and the beginning of the mammal age—needed a more detailed description, and Clemens intended to provide it.

We pulled into Clemens's camp late in the evening and I decided to skip the corned beef and reheated canned cabbage for dinner. The next morning I woke up to an apparition. Inches away from my face were a pair of horse hooves nervously kicking the crunchy, cold silt. I looked up the forelimbs of the horse, which seemed like towering columns against a temple, to see a Diana, a pretty young girl in jeans and a cowboy hat. She squinted down toward me and let out a giggle that sounded more like a baby's gurgle. Then she said, "Howdy!" and rode on to the cabin where Clemens had his maps and his dining table. I couldn't tell whether

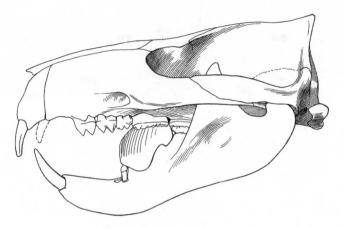

The skull of *Ptilodus*, a multituberculate

the shock of that beautiful young rider or the exhausting drive the night before contributed to my dizziness as I tied my bootlaces. I felt weak and was shivering, but I managed to stand up and shake out my wet bag.

After a greasy breakfast, Clemens guided us to a little amphitheater of sandstone buttes in a semicircle, a spot where he had been working for the last two summers. "Lots of bone here, especially good multi teeth," he said. "Multi" is short for multituberculate, a kind of ancient mammal about the size of a rat, with elongated front incisors and, as the name suggests, curious multiple tubercles or cusps on its molar teeth. These animals were indeed reminiscent of rodents and like them feasted on nuts, seeds, and plant material. But multis were not related to rodents at all. They disappeared from the earth about 48 million years ago, coincidentally around the time rodents began to diversify and probably outcompete multis for food. During the age of the dinosaurs, however, there were no rodents, and multis in Cretaceous fossil beds were as common as rats in a subway tunnel.

I found some big stuff that first day at Hell Creek. Walking up over a knob and then down the cooler, shadier slope I saw the dull frill of a dinosaur, and next to it a projecting horn that curved gracefully to the sky. The tip of the horn was broken off, leaving a rude jagged surface. The horn was golden brown, like the stained

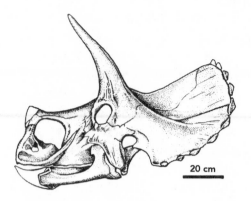

The skull of *Triceratops*

oak of a dining-room chair. "Good one, hmmmm," Clemens smiled. He had wandered by as I kneeled and scraped the silt around the frill with the blade of my knife and brushed the dirt away with a whisk broom. It was *Triceratops*, the horned, frilled dinosaur of great mass, the thunderous, tanklike plant eater depicted in mortal combat with *Tyrannosaurus* in many a museum mural. *Triceratops* is in many ways one of my favorite dinosaurs. It belongs to a group, the ceratopsians, that have the largest skulls of any land animal that has ever lived. *Triceratops* has a skull length of seven feet, and an even more megacephalic relative, *Torosaurus*, of eight feet. In addition, *Triceratops* and its ceratopsian kin had an outrageous pair of powerful jaws with a battery of teeth. These beasts could no doubt have sliced a cedar in a single chomp, although their diet probably consisted of needles, leaves, cones, and other plant material, not whole trees.

"We've got another one closer to the ranch, but this one might be even better. Maybe we can do a little excavating now and take it out when the rest of the crew arrives in a few days," Clemens remarked. I was of course very happy with the prospect of excavating a big dinosaur with my name on it. Even if we had to move out of camp in a few days and head for Wyoming, I might get the pleasure of cleaning the rock from the massive skull of the beast before we encased it in plaster-soaked burlap strips. Alas, this did not happen. The next day, as we got out of the car to walk to the *Triceratops* knob, I started feeling especially dizzy and broke out in a cold sweat. I tried to work for an hour. Clemens looked me over. "You're in pretty bad shape. Let's get back to camp, and I better get you to the doc in Jordan tomorrow." Later I lay delirious with a raging fever in the cool crawlway under the cabin, awaiting evacuation.

The town doctor had long been in retirement. With a national shortage of medical professionals, Jordan was hardly a place to lure a new generation of physicians. He was a pleasant but shaky fellow in his late eighties who prescribed an equally ancient potion of paregoric cut with opium to stop my now-constant diarrhea. The bottle that contained his mysterious powdery chemical looked very familiar; soon I realized it was one of those old opium bottles we used to sort through in the curio shops in Truth or Consequences, New Mexico. For lack of patients and doctors, the only hospital in Jordan had been converted into an old people's rest home. There I lay for two days in an opium-induced stupor, staring at the flies buzzing at the stained, oily lace curtains that fluttered in synchrony with the oak leaves outside. "We got us a real patient," one of the young nurses exclaimed. She kept trying to stuff mashed potatoes and overcooked green beans down my throat and threatened to give me a bath and a rubdown. To these attempts, made with some frequency each day, I protested that I was in the mood for neither. I could hardly leave the can long enough to lie down.

Savage and his crew picked me up on the third day and whisked me south to Miles City, where, feeling a bit better, I made the mistake of scarfing up the first good food that I had been offered in days. But baked ham is not the best remedy for a ravaging gut bacterium, and I relapsed into a worsened condition. A brother of a crew member was a pharmacist in Miles City. He read the label of my opium bottle in horror. "This stuff hasn't been used in thirty years. Here are some pills. Now get him to a real hospital in Billings." As my field mates helped me to the hospital I noticed that my jeans were surprisingly loose. As the scale in the inpatient ward confirmed, I had lost eleven pounds in roughly a week.

The Billings hospital treatment took, and eventually I felt shaky but at least alive, sustained by a diet of bouillon cubes in hot water. Savage was eager to throw my frail carcass in the van and head south to Wyoming. I was feeling a bit like a liability: not only had a curious heatstroke prevented me from working in the field the summer before, but I had now been rendered effectively useless through my own poor judgment of the purity of creek water. Fortunately the scenery mitigated my misery. We were now in the Bighorn Basin, a huge bowl of intricately sculpted sandstone cliffs with rocks that extended from the Paleocene into the Eocene. The Bighorn Basin was famous for great discoveries first made early in the twentieth century and now well represented in fine collections at Yale's Peabody Museum, Princeton (the Princeton collection was given to Yale in 1984 when the trustees decided that paleontology was not worth sustaining with the rise of molecular biology and geophysics), the University of Michigan Museum, and the University of Wyoming. In the Bighorn Basin we spent a few pleasant days with Tom Bown and his University of Wyoming crew.

In subsequent years I have only been a visitor to the Bighorn Basin, dropping in on Tom or others, or joining a field conference, one of those great paleo pageantries where trail stops and fossil scrounging as well as lessons of the local geology are mixed with fire stomps and beer fests. After I got my doctorate from Berkeley, I spent a few days in the Bighorn Basin every year for much of the 1980s. It's a beautiful but intimidating place, a place where a wrong turn up a maze without a good sense of direction, a good map, and adequate water and food can pose a big problem. One evening in the early 1980s, a trio of us pitched camp near the fiery red buttes in the center of the basin. As my friends dozed off, I caught sight of two spidery figures carrying rifles. I shot my flashlight out toward them.

"Don't worry, mister," a faint voice drifted back to me.

The other bleated a very fragile "Help." With the light reflecting the whites of his eyes, he looked like a terrified calf, transfixed in the headlights of an oncoming car.

"Drop your rifles," I yelled, loud enough to stir my mates.

"OK, mister, we'll drop 'em, but please help."

They were two boys, about seventeen, from Worland, a town some fifty miles but many hours away, on the eastern edge of the basin. In their new pickup they had headed across this strange frontier with nothing but two Cokes and a bar of taffy. Their truck eventually got stuck "so bad no one could ever git it out." Without much of a plan, they started hiking west for help, hoping to reach the town of Meeteetse, some unknown miles and direction away. The folly of this venture soon became apparent to them and they headed back toward their truck. But whatever had prevented these unfortunate boys from moving their vehicle had not impeded someone else. No doubt amazed and delighted that the ignition keys were still conveniently dangling from the dashboard, the thieves had efficiently pulled out the truck and headed for the hinterlands.

That was all yesterday morning. For a whole day and night and another day these two boys had stumbled chaotically over myriad drainages. We were the first sign of humanity they had encountered.

"First off, you need some water." I handed them a filled liter bottle.

"No thanks, sir, I've had plenty," said the weaker of the two.

Before I could clarify this mystery, the other gently cradled his bloated stomach. "We've been drinking out of those ponds in the gullies."

Horrified, my mates and I recalled those iron-red mud holes swarming with tadpoles as big as tablespoons. We took those boys back to Worland that night. At the time they were more shaken and wretched than ill, but I suspect that gastric catastrophes were around the corner.

That first year in the Bighorn Basin of Wyoming with Don Savage, Howard Hutchinson, Barbara Waters, and our hosts from Wyoming, we encountered no such strangers, no lost boys with bloated stomachs, no ranchers with stuck trucks, no Basque shepherds, no miners heading for the mountains. The basin is filled with fossils but devoid of people, and we felt comfortably alone among the sand castles. As for me, gastric catastrophes were a thing of the past. I could get back into the work, and I had good luck. When we crossed a saddle between two buttes, Savage's gray Dodge van sunk up to the wheel wells in clay mud. Once we pushed the car to a dry spot, I loped up the hill to piss and spied a shiny nugget with two elongate front teeth on the ground of the intended latrine.

Paramys, an Early Cenozoic rodent, with the body form reconstructed around the skeleton

This was a rodent skull, about 55 million years old, and perhaps the best-preserved and most complete ever found in that region of the Bighorn Basin. In an act of gracious collegiality I offered the skull to our hosts from the University of Wyoming. Howard looked on the ceremony rather darkly and sulked a bit that evening. He dearly wanted that skull for the UC Berkeley collections.

We ended our tour of the bony basins in southwestern Wyoming. Our camp was now a strange deserted ghost town on the Union Pacific Railroad line called Bitter Creek. Seven weeks had passed, and we had long adapted to that trancelike rhythm of sleep, eat, prospect, quarry, eat, sleep again that deadens the brain like the maddeningly constant Wyoming wind. There were only a handful of oil riggers and their families in some disheveled trailers on the edge of the dead town. News of the outside world—Watergate and Nixon's resignation—came in as static on our shortwave radio. I sat in a small gully alone for hours, under a sun that drifted through a mountain range of cumulus, breaking into sandstones and sooty coal seams, looking for a precious fossil, perhaps a primate skull—

"an ancestor of man's ancestors," as Savage poetically called them. When the wind died down I could hear a strange whiz that sounded like a ricochet of a rifle bullet: a pronghorn antelope, annoyed with my interloping, farted a warning signal. Home, home on the range, where the deer and the antelope fart, I chuckled to myself.

We did well at Bitter Creek. Savage delighted in the rich concentrations of primates, insectivorans, and other mammals, and Howard was happy with his piles of turtle shells. He would pick up a chaotic jumble of these bits, stuff them in a canvas bag, and, back at camp, sprinkle these shards in a small sandbox and in his leisure try to fit them together. One evening he sensed me staring at him in wonder and glanced up and smiled. "I like jigsaw puzzles."

We had hundreds of mammal teeth, croc bones, turtle bones— bags and bags of them. This was a far cry from the impoverished days in Monument Valley. I even snagged the big mammal of the summer, a six-foot skeleton of *Hyrachyus*, a primitive plant eater that was something like a cross between a tapir and a rhino. Savage launched into the excavation of the beast. "Good work, Mikey John," he said approvingly.

Among the few visitors to our Bitter Creek camp were Doug Lawson and Mike Brett-Surman, two other grad students from Berkeley. Doug and Mike were after dinosaurs, but not just any dinosaurs. Classic locations like Como Bluff, the Red Deer River, Hell Creek, and Dinosaur National Monument still produced notable skeletons or at least skeletal parts of monstrous sauropods, duck-billed hadrosaurs, and horned ceratopsians. But none of these famous spots yielded much in the way of dinosaur eggs or dinosaur young'uns. Such finds are important, because the arrangement of eggs at nest sites and the skeletal structure in dinosaur embryos, hatchlings, and juveniles can tell much about dinosaur behavior, reproduction, growth, and affinities. It was frustrating that the great dinosaur egg sites were far way in Mongolia and China. Doug had a theory, or rather a speculation, that the well-known dinosaur localities of western North America lacked eggs and dinosaur babies because they represented ancient habitats near the shoreline of an inland sea or along huge muddy river deltas. In other words, these

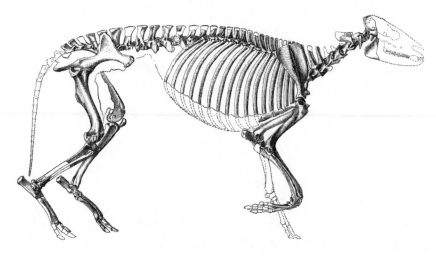

Hyrachyus, as drawn in 1934

dinosaur herding grounds were lowland areas. Doug reasoned that dinosaurs were more likely to have layed their eggs and nurtured their young in the uplands, where steeper, more covered terrain would have afforded more protection from predators and the elements.

Doug and Mike searched all summer in the Evanston Overthrust, a sequence of Cretaceous rocks tilted and well exposed in the hills about a hundred miles west of Bitter Creek and the Washakie Basin. Tired but undaunted, they came to camp for food and brief company. Mike Brett-Surman had other pressing needs. Like Howard, he loved sci-fi and was a Star Trekkie of the first order. Doug complained that he had to stop in a motel every two weeks so that Mike could get enough TV to sustain himself for the next siege on the Evanston Overthrust. Mike cheerfully acknowledge this addiction. "I need a set, gimme a set," he moaned like a tormented junkie.

Doug was not successful in finding upland dinosaurs and egg sites that summer, but he went on to fame for another discovery. Before coming to Berkeley he had studied at the University of Texas, where he spent many seasons in Big Bend Country. There he found the bones of a big flying reptile. He now had the bones of that beast back in the office we shared at the university. One day he

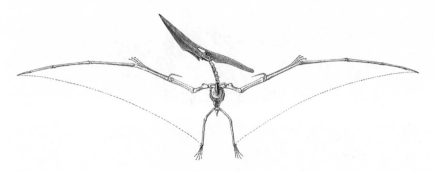

Pteranodon, a pterosaur

started chuckling and called me over. He showed me the humerus of his Texas creature laid out on his worktable next to the same bone of the then largest known flying reptile, *Pteranodon*, a soarer with a twenty-three-foot wingspan. "Wow!" I gasped, as Doug just kept laughing, a little giddy with the astounding observation he had just made. The Texas humerus was more than twice as massive as its *Pteranodon* counterpart. Although the Big Bend specimen was not complete, enough of the skeleton remained to enable Doug to estimate the wingspan of the monster at an incredible fifty feet. Doug had clearly discovered the earth's largest known flying creature. He named the animal *Quetzalcoatlus northrupii*, after the mythical bird of Aztec lore and the inventor of the strange aircraft called the flying wing. A sketch of the "Quetzi" soon graced the cover of *Science*, which along with the London-based journal *Nature* was the most important publication in the scientific community. This was an almost unheard-of accomplishment for a graduate student.

Doug's hunch about the probable presence of dinosaur nests and juveniles in the Rocky Mountain region was also eventually substantiated, but not by him. In the 1980s, long after Doug had decided to apply his brilliant mathematical and computer skills to a job as a well-paid oil geologist, Jack Horner and his team found rich nesting grounds of large groups of hadrosaurian (duck-billed) dinosaurs called *Maiosaurus*. The name translates as "good mother," and skeletons of parents, hatchlings, and juveniles, as well as fossil eggs, were found in a fabulous site known as Egg Mountain. Since those discoveries, Horner and others have found other nesting sites in the Rocky Mountain region rivaling discoveries of dinosaur rookeries made on other continents.

Nine weeks passed, and the late August wind of Bitter Creek had the faintest bite of the Arctic to it. We were even caught in a snowstorm in the gray hills east of town. After a miserable day of prospecting in the cold and sleet, we struggled with shivering arms to push Savage's van out of the frozen mud. Unlike the resolute field team hired by O. C. Marsh to work Como Bluff nearly one hundred years before, we were not disposed to carry on our work in blowing snowstorms. I stepped out into the evening chill and enjoyed the alpenglow on the Cathedral Bluffs for the last time that long summer.

15

. . .

The Curious Beasts of Old Baja

My first summer in Montana and Wyoming was almost my last. I couldn't see fettering myself to a gully of sandstone, extracting every skeleton I might find there, and trying to draw from that a complete picture of the ecosystem with all its carnival of the animals from the Cretaceous, Paleocene, or Eocene. My work in San Diego had prepared me for this so-called faunal project, a fine undertaking on a multiyear scale, but it wasn't the kind of thing one might tackle in the short span of a doctoral program. Berkeley nonetheless was notorious for putting pressure on students to do such things, and the pressure was clearly on me. When I got back from that inaugural trip, Clemens and Savage kept asking me, Where would you like to set up camp and quarry? Some dinosaur badland in Montana? Crazy Mountain Field? A pocket full of fossils below the beautiful Cathedral Bluffs?

My answer, to their disappointment, was none of the above.

Instead, I worked on a collection of very archaic mammals called leptictids, long-snouted, sharp-toothed insectivorans that started diversifying in the Paleocene soon after the dinosaur extinction event. Leptictids are beautifully preserved in skulls and skeletons found in collections at the American Museum in New York, Yale, Harvard, Princeton, the Smithsonian, and other institutions. These fine specimens represent taxa (species) that ranged from the Paleocene through the Eocene and Oligocene, when the last leptictids were known.

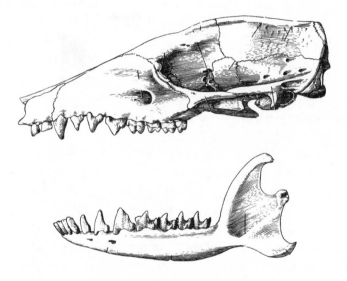

A leptictid skull, as studied in 1986

The leptictids were an attractive group to work on because of this wealth of material and their evolutionary importance. People who studied fossil mammals, especially if they worked field sites, had a tendency to restrict their interest to teeth. This seemed logical, because hard, durable, enamel-armored teeth were by far the most abundantly preserved and reliable indicator of what mammals lived in which given place millions of years ago. But paleomammalogists seemed to be obsessed with teeth to the point where they ignored the more elusive and complex parts of the skulls and skeletons. Specialists in fish and reptiles looked upon this predilection with amusement, even criticized it. Al Romer, the famous

expert on Permian and Triassic synapsids who had been Peter
Vaughn's mentor, once wrote sarcastically that specialists in fossil
mammals viewed evolution as a process wherein teeth copulate
with teeth and beget more teeth. It seemed to me that studying a
group of mammals that was represented by more of the anatomy
than the teeth would be good training for work on a variety of
vertebrate groups, whether mammal, dinosaur, or fish.

During my graduate days, I had in fact a rather schizophrenic
approach to research, moving from one area to another in search of
a perfect problem. With all this meandering, I managed to write articles on multituberculate mammals from the Hell Creek, freshwater fishes from South America and Africa, and even clams and snails
from the Eocene marine rocks of San Diego County. Among the
various topics I touched on, leptictid mammals were good for the
kind of lengthy monographic work required for a doctoral dissertation, since they held a key to many questions concerning the early
evolutionary radiation of placental mammals.

My strategy proved opportune for another reason. Jay Lillegraven was offered a professorship at the University of Wyoming
and I won a competition for his vacated position at San Diego
State. This post was not my ultimate nirvana—the teaching loads
were oppressive—but at San Diego I had good colleagues and
some good students, and the desert was within a few hours' drive of
the city limits. As it turned out, my stint at San Diego State was
productive, but not because of fieldwork. The heavy teaching loads
forced me to use my summers for desk and lab research, and I
didn't even get back to Wyoming until I led my first very small
field foray in the summer of 1982. That same year I permanently
left San Diego because I was unexpectedly confronted with the
opportunity of a paleontological lifetime—a curatorship position at
the American Museum of Natural History in New York. Ironically,
one of the first things I did at the museum was to start planning a
field program that included exploration of a region just south of
San Diego, the great barren peninsula of Baja California.

Actually my interests in Baja had been drawn out many years before, when I was still a graduate student working with Lillegraven

at San Diego State. During a visit from Savage, Lillegraven decided to escort a small group of us across the border to an area of Tertiary rocks called the La Mision locality, which is in the hill country wedged between the Sierra de Juárez and the Pacific. It was summer when we went, and it was much hotter in Baja than anything I had experienced since those sizzling days in the canyons of southern New Mexico. I was, however, entranced with the place. Once we passed the polluted sprawl of Tijuana, we continued on Highway 1 along the coast. I felt the bracing mix of coastal air and the scorched aridity of a waterless plateau. The road was unadorned with the bric-a-brac of southern California civilization—no McDonald's or Carl's Junior's blocked the view of blue sea lapping deserted beaches. I had never been south of Tijuana, and everything seemed strange and new for a land only a Sunday's drive from San Diego. Rickety trucks fumed and sputtered but persevered up steep grades of the highway. We were passed by a red pickup whose truck bed was loaded with eight men sprawled about laughing and singing. One of them tossed out an empty can of Tecate beer that bounced on the road in front of us, its metallic red coating reflecting in the sunlight like a Christmas ornament.

The Baja highway was lined with fruit stands where smiling people in bright clothes stood as straight as sentinels by pyramids of watermelons. Children were everywhere; they walked deliberately down the road for miles between towns, often toting a big sack of rice or beans or a chicken carcass. When they played, they played with ingenuity, using an old rusted bike rim as a hula hoop, or crossing swords with two bamboo sticks. We could not really feel a connection to the daily hardships of these poverty-stricken children and their parents, but I could see the signs of misery—a baby wailing in the spindly arms of a dirty little girl in a shredded blouse who was hardly older than the baby herself, a crippled old man with one eye brushing flies from his lips as he savored some tequila from a wide-mouthed jar. This was a different Mexico from the southern California re-creations of quaint villages, piñatas, and strolling mariachi players. It was also different from the almost monolithic urban squalor of the shantytowns around Tijuana. Baja

California shocked the senses with sights that were both uplifting and, around the next bend, tragic. It was at least more real than any mall town in the lavish, commodity-ridden metropolis north of the border.

We turned off the highway and struggled up a twisting road in a canyon with gigantic mesquite trees that filled the spaces under dusty sycamores and oaks. Near the head of the canyon I was surprised to see a town peeking through the mesquite. Its small, dirt-stained adobe houses loomed suddenly in front of us as we came around a blind curve. The place was invisible from a few hundred yards away. I imagined that Pancho Villa and other Mexican revolutionaries and banditos must have taken refuge in towns like these. "The outcrops are not far from this place," Lillegraven said. The town was devoid of commerce save for a devastated adobe structure with a thatched roof that looked like a bad perm. "That's our emporium for food and beer," Lillegraven mocked.

Our jeep crawled up the rampart that protected the mesquite forest and its odd little settlement and arrived at a brown plateau covered with grass and scrub that seem dried to paper in the sun. Only a few sand ridges in low gullies broke up this flat topography. There we spent the day, picking up fossils along the white strands of what was a beach 20 million years ago. This was Vieja Baja, Old Baja of the Miocene epoch. The sands were once part of a shoreline along a

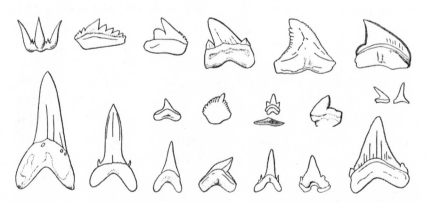

Teeth of different Cretaceous and Cenozoic sharks

bay of shallow water on the western edge of an ancient Baja California. Like other Miocene embayments farther north, in places like Shark's Tooth Hill near Bakersfield, La Mision showed an interesting mix of animals. The spool-like vertebrae and dense ribs of whales and the triangular teeth of monstrous sharks jostled with bones of mammals—not just seals and sea lions, but also dogs, camels, and horses that would have come down to the shore to feed, frolic, and die.

It was entertaining to walk along these outcrops, picking up shiny shark teeth—some real prizes—or brushing the sand away from the bleached white jaw of a fossil seal. Soon I found I had drifted far to the north of my companions and could no longer see them against the rusty, scrub-covered terrain. I closed my eyes and envisioned a long crescent beach, stretching from a distant volcano anchored offshore. From my vantage point I could see a pod of Miocene whales spouting only a few hundred yards away and a pack of long slinky hyenalike forms chasing a frail young camel.

This reverie was interrupted by an intruder. *"Señor?"* I heard a high-pitched voice over my right shoulder. Walking toward me along a range fence a few feet from where I crouched was a cowboy in a broad sombrero and dirty sheepskin chaps. I noticed that he also wore an elaborate silver-inlaid leather belt, well stocked with ammunition and supporting a holster that opened around the sweat-polished ivory handle of a pistol. He flung his legs over the barbed wire without breaking his stride and stood directly in front of me, not more than a yard away. In the distance his horse stomped and snorted, swatting flies off its chestnut flanks with an unruly, thistle-infested tail.

"Cigarettes?" The cowboy cocked his head to one side in a peculiar fashion, as a child in Tijuana begs for candy or money. When he smiled, then suddenly yawned, his mouth looked like a deep pit veined with silver and gold, broken by the blackness of cavities, rotting gums, and missing teeth. "Cigarettes?" he repeated, this time with more earnestness and a tinge of threat.

I shook my head and smiled. "Sorry, *no tengo*," was my awkward bilingual response. He just stood there, smiling and squinting at me.

He looked up at the sky and turned his head east toward the cauliflower of a thundercloud that had been building all day some eighty miles away over the Gulf of California. For hours I had been beckoning that cloud to come hither and bring a little relief from the heat, but it had simply gone on constructing itself into an atmospheric Mount Everest without budging.

"Mucho calor," I said. I thought maybe talk of the weather would divert him, but he simply stared at me, scanning my long hair and making a whistling sound between the sharp edges of his remaining front incisors.

After a very long minute or two he muttered, *"Eheea,"* strode back to his horse, and rode off.

This was not my last encounter with the stranger. As a gesture of dubious reward for the toil of the day, Lillegraven took us all to the adobe tavern in the little canyon village for refreshment. Jay had warned us that he had heard the place was "pretty rough." Upon entering, we were convinced that Lillegraven had only fortified his reputation for understatement. We did not so much see the gloomy bar as smell it; the stench of urine rivaled that of any bat cave I had explored as a teenager. Adapting to the dark, we could see a knurled wooden bar with a few half-empty bottles of tequila propped precariously on a slanted shelf. The floor, what there was of the floor, was covered with broken glass. What little light entered this lair seemed to come from cracks in the floor below and not from breaks in the thatch above, most of it emanating from a gaping hole in the floorboards where a huge piece of granite projected into the room like a looming island. A couple of drunken cowboys were leaning on the granite, urinating copiously. Beyond this centerpiece, I could make out a familiar form in dirty sheepskin chaps, holding a bottle of tequila and leaning on two prostitutes. The women were laughing as they flicked flies off the cowboy's nose.

We slunk surreptitiously to the bar, where soon the cowboy, tired of his female company, staggered toward us. He could barely walk, and his spurs kept catching in the cracks between the floorboards, nearly bringing him down. When he managed to get to me, he drew his gun and waggled it in the air. He kept squinting and then opening his eyes very wide as if he were trying to get me in focus or catch the faint memory of my image. Then suddenly he remembered. "Cigarette?" he asked sarcastically. He smiled and resumed shaking his gun at me.

Savage, always the watchdog, whistled. The cowboy turned his head; Savage was waving a bottle of beer at him in a cordial way. Seeing this respectable older man with well-groomed hair, the cowboy put his gun in his holster and practically fell on Savage in conviviality. Savage did not know Spanish but he had spent enough time in Italy to be able to butcher Italian the same way he violated French. *"Bueno sera, señore, prego, prego,"* he said. The cowboy hunched and tilted his head, trying to decipher this strange Spanish dialect. Savage signaled me behind his back with a wave of his right hand toward the door. He then left a wad of pesos on the bar and was hot on my heels. The others were already in the jeep, having deserted the bar almost as soon as they had entered it. "Nice place, Jay," we all said.

Fortunately nothing about the fossils of the La Mision locality ever induced me to enjoy the nearby village conveniences again. There were other places in Baja, however, that greatly intrigued me. One of these was an extensive set of Cretaceous rocks near the coast far south of La Mision, 225 miles from the border. These beds were exposed as tawny bluffs along a broad river valley, threaded by a trickle from the Sierra San Pedro Mártir, that opened like the maw of a great white shark into the Pacific. Highway 1 cut through this river valley near the small village of El Rosario. Here various crews from the Los Angeles County Museum and others had managed to find rather chunky but impressive dinosaur bones. The most abundant of the dinosaurs was *Hypacrosaurus*, a monstrous duckbill dinosaur, or hadrosaur, which rivaled in size the largest hadrosaurs from the Cretaceous rocks of the Rocky Mountain states. One ischium (the back section of the pelvis) of *Hypacrosaurus* was more than four feet long. A single spine coming off the tail vertebra of one of these beasts was fully two feet. The crews jacketed these fossils in thick plaster monoliths reinforced with wood struts or frames, then dragged this tonnage down the cactus-covered hills in the Baja heat.

A number of fragmentary bones of other kinds of dinosaurs, in addition to *Hypacrosaurus*, were recognized. These included large teeth with serrations that represented tyrannosaurids and platelike dermal bones that belonged to the armored, club-tailed anky-losaurs. Also very appealing was one small marsupial-like mammal called *Gallolestes* in a paper by Jay Lillegraven after the formal name given to these Cretaceous rocks, the El Gallo Formation. The accu-mulated remains of dinosaurs, mammals, and other vertebrates from the El Gallo Formation were not impressively complete—they

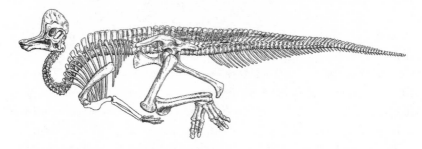

Corythosaurus, a hadrosaur related to *Hypacrosaurus*, from Baja California, as drawn in 1916

would hardly make for a diverting museum display. Yet this assem-blage, even in its bare, broken bones, had some interesting charac-teristics. In a paper published in *Science* in 1967, Bill Morris from Occidental College, the leader of the expedition, had noted that the El Gallo fauna compared favorably with animals from the Up-per Edmonton Formation in Alberta, Canada. The Edmonton se-quence included famous dinosaur beds worked many decades before by the flamboyant Barnum Brown and his American Mu-seum of Natural History crews. It was reliably dated by radioactive potassium (which decays to argon) at 66 to 68 million years before the present, an interval generally regarded by paleontologists as the latest Cretaceous. The El Gallo fauna might also be regarded as of the same age, if one based the estimate on the dinosaurs. However, marine rocks that interfingered with the El Gallo Formation sug-gested an earlier age, about 72 to 73 million years before the pres-ent. Morris thus argued that certain members of the Baja fauna, such as *Hypacrosaurus*, had been restricted to the Pacific margin of the continent during much of the Late Cretaceous, and only in its final phase did they disperse to the Canadian region. The fact that this Baja assemblage completely lacked ceratopsians, like *Triceratops*, which were very common throughout the Rocky Mountain region during the same interval, enhanced its distinctive flavor. It suggested that the Baja fauna might have been well isolated from other dinosaur communities by geographic barriers, perhaps extensive mountain ranges.

Other evidence of the Baja beasts came trickling in. Ralph Molnar wrote a paper published in 1974 describing a fragmentary skull, pelvis, and foot from a new theropod, about two-thirds the size of *Tyrannosaurus*, found in a rock unit just below the El Gallo Formation. This form, which Molnar named *Labocania*, had intriguing similarities to some of the carnivorous dinosaurs from Central Asia. Together with the dinosaurs and mammals from the El Gallo Formation reported on earlier, it made the prospects of discovery in the El Rosario Badlands even more enticing. The theory of geographic isolation of the Baja dinosaur community required testing, and the test could be accomplished only if a much better sample of dinosaurs and other vertebrates were recovered. I accordingly expressed interest in the area, but others cautioned that the rocks were excruciatingly stingy. Even the great collector Harley Garbani, who had found the best dinosaur and mammal specimens at El Rosario and some of the best at Hell Creek, working with Bill Clemens, remarked despondently on his many fruitless days in Baja by the beautiful Pacific.

Toward the end of my faculty term at San Diego State, in the early 1980s, I conducted a couple of sorties into El Rosario. We scoured canyons and cliffs and found some impressively large, stumpy dinosaur limbs and some petrified wood. But the sober predictions of a lack of bone in the El Gallo were well substantiated. The yield of bone in this area was unusually bad, about as bad as I had ever seen. El Rosario made Vaughn's impoverished fossil sites in Monument Valley look like the bone orgy at La Brea. After a few days of vain searching we stood in the clammy fog of an empty beach strewn with surf-polished rocks and driftwood, trying to start a fire from some soggy oak logs that had been dropped mysteriously by the highway. I knew at that moment that I could not justify launching an expedition all the way from New York just to prospect these barren bluffs. In subsequent years, other field parties have had a similar lack of good fortune.

Yet it seemed a shame to turn away from the paleontological opportunities in this lovely desolate peninsula. Even farther south than El Rosario, another 200 miles below the border, was an isolated spot called Occidental Buttes deep within the blazing Vizcaíno Desert. Here in 1966, crews from the Instituto de Geología in Mexico City, the Los Angeles County Natural History Museum, and Occidental College had found a puny sample, only four specimens, but these included the remains of an archaic piglike form called a pantodont, a group that first appeared in the middle Paleocene, about 60 million years ago, and an intriguing occurrence of the most primitive known horse, a three-toed beast known as *Hyracotherium*. On the basis of this small sample, Bill Morris wrote a paper published in *Science* claiming that the Baja fauna was late Paleocene in age. This was a problem because *Hyracotherium* was not known to occur elsewhere before the early Eocene. Morris proposed that the Baja occurrence of *Hyracotherium* suggested that horses might have evolved in the south in the Paleocene and migrated north to the core of the continent in the early Eocene. This theory about the origin and migration of horses, one of the most important mammalian fossil groups for demonstrating evolutionary changes over long periods of time, attracted a lot of attention.

I was very curious about the sample and conclusions regarding Occidental Buttes. Were these Eocene mammals really like those found in many parts of the interior of western North America? This seemed rather odd, because conventional wisdom had it that Baja California during the Eocene was much farther south and opposite a subtropical or tropical part of Mexico and Central America. Baja is really a piece of errant crust that floated northward along with the rest of California lying west of the Gulf of California and the San Andreas Fault. The gulf and the fault extending from its apex are just cracks in the crust, making a tremendously elongate boundary between two crustal plates that separated with the opening of the eastern Pacific Ocean some tens of millions of years ago. One might expect therefore that the Eocene animals of

Baja were very different from those that populated the basins of the Rocky Mountain states much farther inland.

Yet Morris might be right: southern Mexico and Central America could have been a staging area for the mammalian evolution of groups such as horses. But to test this hypothesis would require more fossil evidence. A more complete sample should either show a high number of primitive species, typical of Paleocene faunas in other parts of North America, or complement *Hyracotherium* with a high proportion of mammals typical of early Eocene communities elsewhere. The latter alternative would simply demonstrate that mammalian faunas of the early Eocene, unlike mammalian faunas at many other times including the present, were the same throughout the North American continent. Another kind of evidence could be highly useful too. Analysis of radioactive elements in volcanic or other igneous rocks above, below, or within the fossil-bearing rocks, or a measure of the magnetism within the fossil-bearing rocks themselves, could provide independent evidence of the age of the fauna.

Throughout the sleety winter of 1983, my first year in New York, I couldn't shake my fascination with that motley assortment of fossils representing the Eocene beasts of California Sur. However, my interest in Baja seemed more of a distraction than a wise use of time. I had a full plate getting my research going and thinking about future expeditions to the American West and elsewhere. The American Museum had staked claims in many choice areas of Wyoming and Montana, and there were some field funds available to launch a project soon. Nonetheless, when spring came to New York with its crocuses and yellow-green willows, I thought about a much more magnificent floral display in the spring and early summer of a California desert. And the place had fossils too. In the first week of May I assembled my spare collection of field gear—a rock hammer, a tube of Baja maps, sample bags, some plaster bandages, rattlesnake-proof hiking boots, and vintage, fluorescent orange California sunblock—and flew back to San Diego en route to Occidental Buttes.

16

. . .

Baking Below the Buttes

The 1983 Baja Paleontological Expedition was very small. Besides myself, it included only the late Richard Estes, a former colleague of mine at San Diego State and an expert on fossil lizards and amphibians; John Flynn, a recent graduate and a specialist in geology from the American Museum–Columbia University Ph.D. program; and Andy Wyss, one of my students from San Diego State, who intended to transfer to New York to work with me on his Ph.D.

We left San Diego on a cool, breezy May 9 in San Diego State University van number 77 and made the marathon drive south, hoping to get far enough to reach Occidental Buttes on the subsequent day. By this time I was somewhat accustomed to driving old Highway 1, but the trek is always an adventure to rival the Baja 500, the classic auto race. The brochure on Mexican auto insurance had some foreboding statements: "In the event of an accident with another vehicle, the police will probably arrest you first and then investigate the accident. We will try to keep you out of jail." The police, who were infamously corruptible, had plenty to keep them busy. On my first visits to Tijuana, I had been amazed to see vehicles careening into each other, every traffic light on Avenida Revolución being nonfunctional. But this game of bumper cars paled in comparison to spectacles farther on down the road.

North of Ensenada we saw on the side of the road a semitruck with its cab rolled over, its trailer completely collapsed like a flattened cardboard box. A group of men were casually trying to reconstitute the trailer using twenty-foot pry bars to push up the sides. In the hills around Santo Tomás, fifty miles farther on, we were crawling up a grade in a line of twenty vehicles nudging behind a crippled semitruck carrying a load of fish fertilizer. Suddenly a black Cadillac convertible full of dudes in sunglasses, black shirts, and bolo ties came up from behind and passed the whole lot of us, then spilled into a ditch flanking the shoulder of the road in reaction to the oncoming horde of traffic. As we passed the Cadillac we saw the men laughing and drinking beer, like the two older, rather fat men we encountered a little farther along, who squatted under their burning car throwing sand under the engine with one hand and holding beer cans in the other.

One of the most notorious stretches of the highway was the long five-mile winding grade into El Rosario. Here accidents were so frequent that the bottom of the canyons on either side of the road looked like demolition yards, with carcasses of cars filling the dry arroyos. On this trip we encountered a large commotion. A semi that had slipped off the road down a deep ravine was being towed out at the end of a monstrous cable hitched to the back of a truck. But the towing beast appeared rather ineffective; it spewed black clouds of spent diesel to no avail, and the rescue crew looked on, dumbfounded, as the front wheels of the tow truck started to lift off the ground. All that kept this truck from joining the others at the canyon bottom was a huge pile of rocks behind the rear wheels, where a cluster of men with pry bars were again doing something useless.

When we reached the bottom of the grade at the junction with

El Rosario, the veritable center of town, we found we had just missed another spectacular accident. At the crossroads in this small village, the screaming 8 percent grade of Highway 1 makes a ludicrous left turn as the road continues east and then jogs south to cross a broad wash. A truck had missed this turn and gone right through an adobe edifice unwisely located there. Apparently there were no casualties, but the destroyed house was adjacent to the town jail, a structure about the size of an outhouse. A wretched prisoner with his hands on the bars of his tiny window was screaming and cursing, begging to be freed from an incarceration that seemed an unintentional death sentence.

We stopped in El Rosario at our favorite general store, located safely a few hundred yards from the deadly turn. There we bought some of our usual staples for a field trip in Baja—Gamesa crackers, fresh tortillas (good for only a day in the heat), Herdez canned green salsa (very hot), refried beans, onions, garlic, fresh and dried chiles, and rice—all for a measly 1,060 pesos, about $3.50.

It was now 3:30 p.m. and we were eager to get through another treacherous section of the winding road south of El Rosario before making camp. *Gallo* in the name El Gallo Formation is the Mexican word for chicken. Coincidentally, or not, this stretch of the road was thick with lumbering trucks pulling twin trailers loaded down with chickens in piles of shifting cages. Following one of these sluggish vehicles uphill as it emanated raw diesel smoke mixed with the smell of chicken shit was truly agonizing. But far worse was confronting one of these "Gallo Express" trucks as it screamed down a grade taking up much of the opposite lane. To make matters worse, the highway traffic increased at dusk, since the big semis with their riddled radiators are better able to survive in the cool of the night. The drivers turn off their headlights as a statement of both machismo and energy conservation necessitated by their half-dead batteries. These are far less than ideal driving conditions. As Mark Norell, one of the team members in the Baja 1984 and 1985 expeditions, once remarked, our last moment on earth will probably be marked by an image of a dark truck cab coming at us dead-on, with a flash of gold teeth and a tequila bottle on the dashboard.

Just south of the El Rosario wash is another spectacle, happily not related to automotive disasters. Among the clumps of coastal ice plant and small creosote bushes a few of the weird, rocket-shaped cactus, the *Idria*, or boojum, pop up. Soon these boojum proliferate and the hills bristle with green candlesticks. A few miles farther along, as the highway makes sweeping switchbacks, the boojum are joined by giant cardon cacti and the twisted, gray-barked tree with golden leaves called the elephant tree. The stunning botanical mix becomes more dramatic farther south, near the small resort stop of Cataviña, where boojum, cardon, yucca, cholla, and barrel cactus were in bloom, along with stunted, bonsai-quality elephant trees in grottos among the huge granite boulders. There is probably no lovelier cactus garden on earth. Cataviña was then a fitting destination to end a ten-hour drive, and we slept on the soft sand of a wash between the boojums and the boulders on that cool night.

The next morning, May 10, was typical for the late spring of Baja. Cataviña lies some forty miles east of the ocean, but the maritime weather is aggressive enough to make a nocturnal creep inland. We arose at 5:30 a.m. to find ourselves engulfed by a dense fog. The cactus garden had become a cloud forest; the boojum were enshrouded and druidlike in the mist. Our sleeping bags, food sacks, and camera cases were completely soaked. We warmed our shivering bodies with hot coffee and delicious *huevos rancheros* at the El Presidente Hotel and then headed south toward the Occidental Buttes.

The remaining challenge was actually to find our target area. We had U.S.G.S. and Mexican topographical maps, but it was unclear whether any kind of road or jeep trail would lead us off the highway to a place where we could reach the Buttes or at least see them clearly. I knew that our van was not likely to be up to any road conditions much worse than the 420 miles of potholed highway we had just negotiated. From what some earlier prospectors had told me the Buttes were isolated and easy to miss, and getting to them would be an adventure. This turned out to be absurdly untrue. We rounded a bend in the highway and could look directly southwest to an imposing pyramid of soft red sandstone capped by a crown of pebbles and rocks known to geologists as a conglomerate. This was a butte and it had to have fossils.

The other inaccuracy pertained to the name of the place. Gringo paleontologists had named this butte and the smaller one northwest of it Occidental Buttes, perhaps in honor of the home base for the expedition, Occidental College—and this was the name picked up on the U.S.G.S. maps. But this rather bland identification obscured the original, much more descriptive name on the Mexican topographical maps: Lomas Las Tetas de Cabra, or Hills of the Goat's Teats. Henceforth the Mexican name for the place became the official name in all our scientific publications.

We knew from the accounts of the 1966 expeditions that bones were very rare at Las Tetas. But one of the greatest pleasures in paleontology is to be the first to return to a spot in more than a decade or two and to find that the work of wind and those rare Baja rains has scraped off some of the fossil-bearing surface. We did remarkably well even on our first afternoon. Soon after arriving, Andy found the tooth of *Meniscotherium*, a small, cloven-hoofed animal distantly related to modern deer, antelope, and bovids. *Meniscotherium* was, interestingly enough, also very typical of early Eocene faunas of the North American interior. A couple of hours later I had the great luck to find that elusive primitive horse *Hyracotherium*, represented by a nice lower jaw with two molars. We also found some lizard vertebrae and several toothless jaws.

Prospecting the steep slopes of the twin Tetas was good exercise. More difficult were traverses across ravines to reach a likely-looking outcrop on the opposite side. Many of these ravines were choked with the evolutionary acme of the thorn garden—devil cactus, teddy bear cholla, brittle brush, barrel cactus, and more. We called such an area Pecos Bill Land, PBL, after the children's-book cowboy hero who hacks through such thorny country on his trusty steed along the Rio Grande. Sometimes the tedium of hiking and prospecting up the slopes or through PBL was interrupted by childish diversions. We would face each other on a narrow ridge and try to bully or push each other off down a steep slope on either side. We named this particular sport Ridge Chicken, and it was complemented by Chasm of Death, a competition to determine who could jump across the widest gap between two ridges flanking a yawning crevasse, or Slope of Death, a form of extreme skiing in hiking boots down a ball-bearing slope. With his Swiss genes and climbing experience, Andy Wyss was the best at these sports; with his extra bulk, John Flynn was the worst; and I came out somewhere in the middle. Our senior colleague, Richard Estes, demurred and tried to understand such childish behavior.

The next day proved equally productive, with numerous finds of bone fragments and several key specimens. I found the rear end—the pelvis, tail, and hind limbs and foot—of a primitive herbivorous animal, probably *Hyopsodus*. Perhaps most interesting was John's discovery: a nice jaw with two premolars and two molars of an opossumlike creature. We also found a jaw of a primitive carnivore-like form known as a creodont.

These were very satisfying results. The specimens were not exceptionally well preserved or imposing, but in two days we had already tripled the diversity of vertebrate animals recorded by the 1966 expedition. Moreover, only the *Hyracotherium* was of a taxon that had been reported previously. All the other animals we retrieved—the opossumlike marsupial *Hyopsodus*, the rodent *Meniscotherium*, the varanid lizards, and a salamander—were new. It is sometimes surprising how little paleontologists need to find to say something meaningful. A puny assemblage like this in the Bighorn or Washakie basins would hardly warrant a second glance, but in the most isolated and southerly spot in North America where Early Tertiary land vertebrates had been discovered, every fragment of bone or tooth was worth cherishing and closely scrutinizing.

That first reconnaissance to Las Tetas was meant to be brief. We returned to San Diego State on May 15, only six days after we had left it. For a few days before returning to New York I enjoyed studying our new discoveries. Then Estes got some disturbing news: Ishmael Ferrusquia, a paleontologist from the Museo del Instituto de Geología in Mexico City, arrived at Las Tetas together with his crew within a day or two of our departure only to find "footprints all over the locality." Rumors blamed us for trespassing. We were innocent of wrongdoing: I had written to Oscar Carrenza, another paleontologist at Mexico City, for permission and my request had been granted. But apparently communication had broken down between Oscar and Ishmael. The problem was resolved in a fair way. We would continue to work Las Tetas and Ishmael would visit us in New York and collaborate with us on a monograph summarizing his findings as well as ours. The fossils themselves, after we prepared, studied, and wrote about them, were ultimately the property of the Mexican government and would be returned to the Museum in Mexico City.

With the support of the National Geographic Society, I returned to Las Tetas in the summers of 1984 and 1985. Again I had a small, highly skilled crew that included Andy Wyss and John Flynn and his student Bob Cipolletti, as well as an American Museum postdoc named Larry Flynn (no relation), another talented San Diego grad student, Mark Norell, an American Museum preparator, Jeanne Kelly, and a Museum grad student, Nancy Olds. They were a relatively young lot, and I, merely in my early thirties, felt a little odd as the senior member and expedition leader. We were indeed a casual, happy, and productive crew.

As the seasons wore on, our habits and profiles seen at close quarters attracted some nicknames. Andy, for example, was called "Action Wear," because of the inexpensive Sears Action Wear clothes he preferred. Andy had a penchant for finding the best deals in anything—his favorite restaurant in New York was a modest establishment called The Food Shop—and his Sears clothing seemed amazingly clean and durable, resistant to the nastiest thorns in any Pecos Bill Land. Larry Flynn was nicknamed "The Alien" because he could recall perfectly the lyrics of every bad television show tune from the 1970s and 1980s. We reasoned that anyone with that combination of encyclopedic knowledge and bad taste must have been programmed on his mother ship to "blend in with earthlings." Television, particularly the series *Gilligan's Island*, inspired a number of other nicknames. I was "The Professor," John "The Skipper," and Mark "Gilligan" or "Li'l Buddy," an appellation he deeply hated. Mark had a habit of relentlessly teasing John. When John reached the threshold of tolerance he would say, "Aw c'mon, Li'l Buddy, let's be friends," and cuff Mark on the back of the head.

With the support of the American Museum and outside funding, I was better equipped for a more elaborate venture in 1984. In early June we drove cross-country in a new red Ford F-250 pickup truck and a big tan Chevy Surburban, nicknamed Blondie. In San Diego we stopped at Ralph's Supermarket and loaded up seven shopping carts. The cashier shouted "Wagon's ho!" as he rang up $650 worth of goods. Maritime air drifting in from the Pacific, only six miles away from Las Tetas, kept the afternoons breezy and the evenings chilly. We sat in our down jackets around a fire stoked with skeletons of dead cardon cactus.

In 1985 I made the mistake of scheduling the expedition for July. I assumed wrongly that the nearby ocean would still mitigate the broiling summer weather typical of the Baja desert. Instead, we sweated under a sun that heated the still air to 105 degrees by 10:30 a.m., all the while staring at a salubrious cloudbank over the ocean that never drifted our way. And the night also failed to bring relief. The cool, dry nocturnal air was usurped by a blanket of warm, humid low pressure flowing west from the Gulf of California, where thunderstorms and even hurricanes were spawned and extinguished before they came within a hundred miles of our camp.

The heat brought with it other inconveniences. Baja has more resident species of rattlesnake than any other region in the world, and we were treated to the frequent sight of this diversity under rocks, below elephant trees, or under piles of driftwood on a rocky beach. Baja also shares with other parts of Mexico, Arizona, and New Mexico the distinction of harboring the world's most deadly scorpion, *Centuroides sculpturatus*, a small but lethal creature that has killed more people in the United States (mostly children and older people) than poisonous snakes. On one of those warm nights I slept outside my bag on the open ground and awoke to a burning sensation on the insides of my thighs. As I brushed my hand down my legs I smashed two young *sculpturatus*, but they had already done their dirty work. Fortunately, there was a little ice to apply to the swollen, burning puncture sites. We knew we could drive the sixty miles to Guerrero Negro and radio a plane for evacuation, but this did not seem necessary, given the decent state of my health and the limited poison injected by young scorpions. Nonetheless, I spent a miserable and rather anxiety-ridden eight hours suffering nausea, fever, shakes, dizziness, and groin pain.

The heat and dryness of Baja became the existential conditions that defined all our strategies and activity patterns. We stored as much food as we could on dry ice bought in San Diego, and feasted on fresh meat and vegetables for a few days. Mark was especially creative in his self-appointed role as camp chef. A sequence of menus included paper-wrapped chicken, *carne asada*, steak *au poivre*, barbecued fresh fish, basmati rice with raisins and cloves, Chinese hot pot, and Mark's specialty, gunpowder chile (actually made with saltpeter and three kinds of chile peppers). We opportunistically consumed the best and freshest food first, with a resolution to tough it out later. So the meals became more predictable as the weeks went on—dried beef with refried beans, spaghetti with clam sauce, or maybe just unadulterated beans, always with a few blazing chilies as garnish. Even in the later weeks, Mark had an ingenious meal plan. One evening he scraped out the heart of an agave cactus, wrapped it in tinfoil, and roasted it in the embers of our campfire to delectable outcome.

The opportunity to dine out tempted us when the repetitive nature of the meals in the late season got us down. Some five miles down Highway 1 was the tiny Café Marisco, which essentially represented the entire town of Rosarito. We had stopped there once and encountered two surfers from Redondo Beach, who told us, "Man, the food is awesome here—you always get sick, though." Our craving for some variety made us ignore this warning. I treated the crew to a feast of lobsters, shrimp, pescado, rice, beans, three different kinds of delicious salsa, and thirteen *cervezas*—all for the measly sum of 12,000 pesos, or about $36. We did get sick, though.

A classier joint required a longer drive. The El Presidente Hotel, some one hundred miles north of our site, offered great food with no evidence of gut trauma, a voluminous swimming pool, and a good excuse for a two-day furlough. We could dine on heavy, dark wood tables draped with festively colored tablecloths under a brick barrel-vaulted ceiling that gracefully blended with the heavy plaster archways and walls. One very pleasant afternoon we shared the dining room with a group of truckers who were pretty smashed and thoroughly enjoying a televised Dodgers baseball game pitched superbly by Mexico's great countryman and hero, Fernando Valenzuela. Suddenly, the set blinked and a stiff-necked man was standing on stage with a pointer next to an anatomical chart. This image came with the announcement "We interrupt this program for the latest update on President Ronald Reagan's colonic surgery." The truckers started cursing and throwing beer bottles against the wall. Several staggered out of the dining room, and soon we could hear the blast of erratic diesel engines. Not a good night to be on the highway facing down Gallo trucks.

In camp we could not indulge in beer like those truckers at the El Presidente. Like everything else, beer had to be carefully rationed. To complicate matters, the clear bottles containing our favorite brand, Corona Extra, photodecomposed and putrefied in the persistent rays of the sun. To transform a bottle of untainted beer from boiling to tepid, we wrapped it in wet burlap and buried it underground. But this indulgence in beer cooling was rare because it required an ample water supply. Water to go with our meal or sustain us out on the outcrop was also very closely managed. We could drive sixty miles south to the town of Guerrero Negro and get water that was ostensibly safe because it was treated with a process known as *electropurificato*. This involved some mysterious

Our fine locality, Lomas Las Tetas de Cabra

wires and tubes stuck in an aquamarine plastic urn full of water. One day I stood outside the "purification building" and noticed that our treated water was being mixed in vats with some runoff that looked suspiciously like sewage. We henceforth treated all water, "electropurified" or not, with unappetizing iodine or bleach.

What little fresh water we had in camp was more than could be found for miles around us in one of the driest spots on earth. And the standing water in our washtubs, the soaked burlap beer coolers, and our own sweat attracted hordes of insects. Under such conditions, a desert can become far more infested with bugs than a steaming tropical jungle. The bees were sometimes so thick that they covered all areas of damp ground and the surfaces of water jugs. It was not uncommon to be awakened from an afternoon siesta with the sound of a buzz in your ear or a violent vibration up your nostril. The insect invasion induced us to keep a smelly trash fire burning at all times. The smoke, the insects, and the makeshift burlap shelters together made for a look of squalor, an unappealing contrast to the red summits of Las Tetas looming over what affectionately came to be known as Camp Tijuana.

Mark was better adapted to the heat than most of us. He kept talking about tanning himself with sunblock formula zero while we sought the shade of our tattered burlap. In the end even he faced his limit. "I think I'll take a sauna," Mark proclaimed once in the heat of the day. He grabbed a quart bottle of beer and entered an incandescent North Face yellow dome tent. The rest of us sat outside it, under the tarp, transfixed by his disappearing act, as if we were waiting for a body to bob up on the surface of the sea. After about half an hour, Mark staggered out of the tent with an empty bottle of beer and almost fell flat into a prickly pear cactus. "It's hot in there!" was all the pain he could share with us.

In such heat, one must adopt a certain pace in desert prospecting. If one is reasonably fit, it is possible to take a vigorous hike in the shadowy hours of dawn and dust during a single weekend outing. But several weeks of this kind of heat weaken rather than fortify most normal humans. A siesta in the heat of the day between the hours of noon and two is not only a relief but a prescription for survival. Likewise, desert hiking requires a realistic notion of the limitations of one's stamina and equipment. On occasion a member of our prospecting party would come back with grotesquely blistered feet. One such exercise forced John Flynn to stay in camp for a couple of days. A desert expedition is only as strong as its weakest member, or rather its weakest member's feet, as we all knew.

Las Tetas, especially during that July 1985 season, was in fact the hottest Hades I had ever experienced. As a local who ran a gas station farther south on Highway 1 exclaimed, "Too much hot, man." Nonetheless, this part of the Baja desert had one redeeming heat breaker. Within six miles was a miraculous deserted beach on an azure Pacific. We soon realized we could rationalize occasional trips to the beach for paleontological purposes. We shoveled out several of our sites around the buttes and loaded the rock into burlap sacks for screen washing. It was a delight to shake the screen boxes in the cold, foamy surf. The beach west of Las Tetas stretched from one horizon to the other without a single edifice or sign of human habitation to break the seamless shoreline. I thought what a precious piece of property this would make in California Norte, where civilization was possible solely because of the technological thievery of water from Owens Valley and the Colorado River.

The Tetas beach was loaded with smooth round stones polished by waves rolling in from some 8,000 miles away without any intervening island to break the force of the swells. There was plenty of nature around. Once, a huge, charcoal-colored sea lion seemed to lie dead on the beach, but when I touched its neck it barked, swung its head violently toward me, tried to take a nip out of my hand, and then slid rapidly into the surf. The only sign of humanity besides ourselves in this Pacific paradise was "the Rock Men," a mysterious crew who came down occasionally in a massive dump truck, loaded up the truck with smooth stones, and carried them off to who knows where.

We withstood the conditions at Las Tetas long enough to accumulate a respectable but fragmentary list of fossils and numerous localized sites for these finds. Following tradition, each of these individual fossil sites was named and carefully mapped. Many of the typically rather bland names for sites, such as East Hill Cairn, Windy Gap, and Marsupial Hill, were combined with more color-

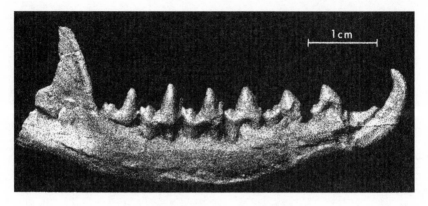

The lower jaw and teeth of *Wyolestes*

ful and more esoteric names, such as Younger World, Li'l Buddy, EEEYI-HA, and Mutant Hill, whose meanings, established through some anecdote or irreverence, were known only to us. We even found and named other sites during one long excursion several hundred miles south on Highway 1. Mark appropriately named a baking outcrop in the hottest part of the desert "White Dwarf." Another locality near the ocean, stuffed with Miocene shark teeth and whale bones, was named "Puberty Blues," after an Australian surf movie we caught on the tube at an El Presidente Hotel.

Those summers of 1984 and 1985 were worth the broil in the heat. We resolutely returned to the same patch of crumbly sandstone repeatedly, mistrusting our thoroughness on earlier occasions. I remembered Vaughn's emphatic exhortations in the canyons of New Mexico to walk and scan every square foot of ground dozens of times. In Baja, the practice had a better payoff. We managed slowly to accumulate an interesting array of fossils, especially more *Meniscotherium* but also fifteen other kinds of mammals representing horses, primitive herbivores called condylarths, early cloven-hoofed artiodactyls, rodents, carnivorans, and a number of odd extinct groups. Among these latter was *Wyolestes*, a long-snouted, sharp-toothed creature.

Wyolestes and its kin, the mesonychids, are fascinating because they are land animals that may be the closest relatives of whales, a transformation now documented in the fossil record with some stunning intermediate forms like the Eocene "walking whale" from Pakistan, *Ambulocetus*. This creature had a body like a whale's but a pair of stunted limbs, which suggests it lived its life somewhere in a limbo between water and land. *Wyolestes* does not show such whalelike characteristics but it does show enough features of the teeth, skull, and skeleton to suggest the kind of animal from which other mesonychids and eventually early whales evolved. This was a very rare mammal in other localities of North America, but oddly enough it was by far our best-preserved species, represented by a good skull part with a full set of upper teeth, lower jaws, and much of a skeleton. Jeanne Kelly found the best and first specimen of *Wyolestes* on a morning in 1984, and the exquisite preservation of the fossil inspired our bone hunting for the rest of that and the next summer.

The fossils would tell us something important about the age of our Baja locality, but we also welcomed evidence for the age of the rocks that had nothing to do with fossils. This meant detecting a signal based on the properties of the rocks themselves, such as their radioactive elements. Unfortunately, the kind of igneous rocks that contain such radioactive elements were either too far above or simply too far away from our rock sequence to be useful. We did, however, have another technique for estimating the age of rocks. John Flynn and his graduate student Bob Cipolletti extracted small, cube-shaped chunks of sandstones at various levels in the rock sequence. Hiking up the slopes of one of the buttes, they measured the thickness of the rock with a makeshift device called a Jacob's staff, which was simply a geologist's compass called a Brunton mounted at the end of a pole. The Brunton not only determines compass direction by pointing toward the north magnetic pole, as any compass does, but also has a dial marked off in 360 degrees with a bubble level, a contraption that allows one to determine the thickness and the tilt of the rock by determining the angle and the height of a particular rock sequence at eye level. At first the procedure using the Jacob's staff seems a bit complicated, especially when you are swatting flies while trying to steady the pole and the compass. But the calculations amount to nothing more than those you make in the first few weeks of high school trigonometry: based on your line of sight to the rock face, the bed thickness (using the length of the Jacob's staff), and the angle of dip of the beds.

The little cubes of sandstone, marked for the longitude and latitude of the exact locations where they were found, their angle of dip, and their position in the rock section, were taken back to the lab at Rutgers University, where John Flynn had nabbed a job as an assistant professor in geology. Here, instead of sampling for radioactive minerals, John and Bob measured the preferred direction of tiny iron-containing grains in the rock. This direction is determined by an amazing physical phenomenon. For some reason, perhaps the interaction of the iron core with the fluid mantle of the planet, a gigantic force field, actually a magnetic field, encircles the earth. Its lines of force converge at the magnetic poles, points coincidentally not too distant from the rotational north pole and south pole of the earth. (This is not the case on certain moons of Neptune and Saturn, where the magnetic poles and rotational poles are widely separated.) We can think of the earth's sphere as having an enormous bar magnet skewering it like a shish kebab. The magnet of course has a positive end and a negative end. The tiny grains in a rock align with the forces of the earth's magnetic field just as iron filings are drawn to the end of a magnet. If we think of the grains as iron filings, we can reconstruct the earth's magnetic field at the time these grains solidified and the rock cooled.

One of the great discoveries in the study of the magnetic properties of ancient rocks was the finding that the earth's magnetic field oscillates through time. In other words, the positive and negative ends of the earth's giant magnet change places periodically. Paleontologists and geologists call the periods of time when the earth's magnetic field had the polarity it has today normal, and periods when it was flipped are called reversed. By the early 1980s, it was clear that the earth's magnetic field had flip-flopped more than one hundred times in the previous 100 million years. Periods of normal and reversed polarity were not always of equal length, and the variation in these intervals offered markers on a kind of magnetic calendar, a timescale that showed stripes of normal periods and reversed periods. In fact, the detection of this magnetic record on the ocean floor in the 1960s was a pivotal clue about movements in the earth's crust, demonstrating plate tectonics and continental drift. The magnetic signals, usually called paleomagnetic signals, showed that the oldest rocks were farthest away from the spreading ridges and the youngest rocks were nearest them. Moreover, the barber-pole striping that could be graphed from these paleomagnetic measurements showed the same pattern on both sides of the ridge. This mosaic vividly showed that the ocean floor was spreading at a uniform rate on either side of the ridge.

With a general assumption that the rock sequence at Lomas Las Tetas de Cabra spanned a stretch from the late Paleocene through the earlier Eocene—about 59 to 54 million years—John and Bob could focus on a time line that seemed most consistent with their paleomagnetic measurements of the Baja rocks. In doing so they identified a reversed event in the sequence, which matched up best with the reversed interval Chron 23 (CR23R) or a short reversed phase within the normal interval Chron 24 (CR24N). These reversed events could be correlated with rocks elsewhere in the world dated both by a rich sequence of fossil marine creatures and by accurate radiometrics. The result was clear—the best paleomagnetic estimate for the age of the Baja mammal community was early Eocene, about 55 million years old.

John and Bob had another paleomagnetic result, this one not so straightforward. The orientation of magnetized grains in ancient rocks not only reveals the state of the earth's magnetic field but also the position of the earth's magnetic poles. The angle of the orientation compared with the surface of the earth could be used to plot a "wander" of these poles through time. However, the poles' shift was partly real and partly apparent. The poles did wobble through time, but not in the radical way recorded in the rocks. The difference in movement can be accounted for by the actual movement of the rocks through time, relative to magnetic north or south, as caused by plate tectonics and continental drift. When they calculated for such movements based on paleomagnetics, John and Bob found that Baja had shifted around a bit but had not drifted radically northward since the Las Tetas rocks had been deposited. This result was truly weird. Other geological evidence was consistent with Baja's having had a much more southerly position and with its having drifted northward several hundred miles since the early Eocene. The data John and Bob mustered seemed sound, but the results continue to be disputed.

Back in New York, our collection of species from Lomas Las Tetas de Cabra, which included amphibians, lizards, snakes, and crocodiles as well as mammals, became a matter of intense interest. The fauna, along with new paleomagnetic results coming out of John's lab, made it possible to resolve the age of the site and make some major observations about vertebrate evolution on the North American continent. The age determination turned out to be decisive. The horse *Hyracotherium*, several of the condylarths, the artiodactyl *Diacodexis*, the meat-eating creodont *Prolimnocyon*, and the tillodont *Esthonyx* were typical of early Eocene (about 55 million years ago) and not late Paleocene (about 58 million years ago) animal communities. In other words, the Baja fauna was younger by about three million years than originally thought. Morris's theory, that horses arose in the south and migrated north into the heart of the continent, was contradicted by this new evidence.

With these new fossils, the Baja fauna looked if anything more like early Eocene faunas elsewhere in North America. In fact it could be compared favorably with those of early Eocene age in New Mexico, Colorado, Wyoming, and Montana. Most notably, a fossil site of similar mammals in the high Arctic, on barren, ice-encrusted Ellesmere Island, had been newly discovered. This showed an extraordinary level of similarity among different habitats. Imagine North America today with the same kinds of animals and plants stretching from the pole to the equator. At present we see a distinctive belt of communities: caribou herds and polar bears in the high tundra, alpine forms like picas and mountain goats in the Rockies, bison and prairie dogs in the grasslands, jaguar and peccaries in the southwestern deserts, and coatimundi and tree sloths in the Central American tropics. Of course, a few species are exceptionally widespread. The mountain lion *Felis concolar* ranges from the Alaskan North Slope above the Arctic Circle to Tierra del Fuego at the southern tip of South America (it claims the record as the farthest-ranging large mammal species). But most mammal habitats are distinctly separated according to latitude, elevation, and climatic conditions.

Our work on the Baja fauna was surprising evidence for a striking uniformity in ecological communities in North America—from the pole nearly to the equator—55 million years ago. Indeed, the paleontological results jibed well with the recently published evidence of fossils from Ellesmere Island and emerging evidence, based on pollen and measurement of oxygen isotopes preserved in fossil marine shells, on the Eocene climate of the earth. Indeed, it became apparent through various studies that the average annual global temperature during the early Eocene hit an all-time high for the last 150 million years. As recognition of the importance of the Las Tetas fauna, in 1984 *Science* published a preliminary analysis written by John Flynn and me. Several other publications, including an American Museum monograph, followed. The combined results yielded a picture of the early Eocene earth as a hothouse that nurtured a Garden of Eden—with dense forest, lakes and rivers, and diverse mammals, birds, amphibians, and reptiles—that has not reappeared over such a wide swath of the earth's surface anytime since.

17

. . .

Whales on Mountaintops

For years I had heard about blind luck, triumph, and fame in pale-ontology. A famous skit staged around the campfire mimicked a fa-mous paleontologist as he twirled in a drunken stupor and literally stumbled over the famous skull of an early humanlike primate. Sto-ries were told about how people found famous dinosaurs in their own backyard or, like Jack Horner, identified bits and pieces of baby dinosaurs in a coffee can in a rock shop. One of the most popular famous tales was about George Olson, a member of the illustrious American Museum of Natural History Central Asiatic Expeditions under the famous Roy Chapman Andrews. In the summer of 1923, Olson tripped and tumbled down a slope at the famous Flaming Cliffs in Mongolia and nearly landed on the first nest of dinosaur eggs known to science.

By the time we had wound up the Baja project, my experiences in paleontology, by comparison, lacked a sudden weird, serendipitous discovery. Our findings at Las Tetas were of value and attracted widespread interest, but they were the result of a rather measured strategy and some stubborn prospecting over the course of three summers. Soon, however, I was to have my own encounter with good luck, an event that was to redefine our field explorations for more than a decade and to open up a whole new region of the world for the study of vertebrate fossils.

This all happened because a flight was delayed at Kennedy Airport in New York. A young man on that flight, Dr. Paul Raty, was a veterinarian on his way from his homeland in far southern Chile to see his relatives in Belgium. On his departure, his uncle, a well-educated man who was mayor of the small town of Chile Chico, told Paul what to do in case his flight was delayed. The uncle noted that a well-known paleontologist, George Gaylord Simpson, had worked at the American Museum of Natural History in New York and in the 1930s had explored the badlands of Patagonia, Argentina, which lay just east of Chile Chico. If Paul found himself delayed for some reason, he should visit the museum in New York and tell whoever now worked there about some of the interesting things the Chileans had found.

One morning in autumn, 1984, the departmental secretary rang me to say that there was a man from Chile who wished to see me; he was at the entrance to the museum. I went down to the great rotunda, the principal entrance to the museum on Central Park West, to greet my visitor. Paul Raty spoke haltingly, in very poor English, explaining that his flight had been delayed for eight hours, that his uncle had sent him as an emissary, and that he wanted to show me some pictures of fossil bones. He was a pleasant, somewhat shy fellow whose smooth skin and well-sculpted, aristocratic European profile hardly suggested his upbringing in a small town on one of the world's rawest frontiers. Only a sunburnt face that seemed alive and healthy compared with the pasty faces of most New Yorkers betrayed his outdoor life. On our way up to my office, he passed, wide-eyed, through the dinosaur and ancient mammal halls, where the great skeletons glared like angry, awakened gods above the throngs of hundreds of clamoring schoolkids. Paul stopped at a mammoth skeleton and pointed to one of the largest vertebrae in the trunk. *"Mismo"*—the same—he said cryptically.

Once in my office, he pulled out some crude color photographs showing large spool-like fossil vertebrae propped up in the grass outside a charming country house. People were standing behind the vertebrae, among them the beaming, gray-bearded mayor. Other photos showed poorly focused close-ups of the vertebrae lying next to rulers, hammers, or other implements used to establish the scale.

"These are vertebrae, probably fossilized," I said, without much emotion.

Paul did not register surprise. He was a vet and he knew his bones.

I looked more closely at the pictures. "They are actually whale vertebrae, big whales."

Paul then gave me a rather puzzled, almost hurt expression, as if I was putting him on.

"Yes," I continued, "they are at least tens of millions of years old. They're well known along the coasts of Argentina and Peru, and I'm not surprised to see them along the Chilean coast too, especially in the north."

Paul was now staring disconsolately at the photos as if he weren't listening to my lecture. He said nothing; he just shook his head. I was ready to escort him out of the office, but he just stood there so expectantly that I had to ask, "Did you find these in Peru or northern Chile?"

He shook his head again, and I led him down the hall to the peaceful Osborn Library. This room held an impressive cache of thousands of reprints and books accumulated by Henry Fairfield Osborn, a paleontologist and president of the museum during the early twentieth century, as well as by many others of his successors. We took chairs next to the stately cherrywood desk with a black leather writing pad that had once belonged to the intrepid, and very competitive, dinosaur hunter Edward Drinker Cope. I brought out a large folio atlas and opened it to Chile, whereupon Paul tentatively dragged his finger down along the spine of the Andes, bringing it to rest on a mountain just south of a large lake in the southern part of the country. The summit was about 10,000 feet in elevation.

"*There?* In those mountains?" I stammered incredulously.

For the first time Paul nodded. He looked serious, almost fearful, as if my disbelief was threatening.

"Are you sure?" I did not want to insult his credibility, but the tale was just too crazy to accept. Fossil whales were indeed common in certain beds along the coasts of Peru and Argentina. But to my knowledge there was no record of a fossil whale beached 10,000 feet in the air on an Andean summit. Obviously this had strong implications for the history of southern South America and one of the earth's tectonic paroxysms, the vertiginous rise of the Andes. Had that episode of rampant mountain building ripped up old coastlines and elevated them to the stratosphere? Moreover, there was virtually nothing known about the inland vertebrate fossil history of Chile. Neighboring Argentina had all the fossils and all the attention.

"We'll be right down," I said. I smiled and shook Paul's hand and led him back to the taxis along Central Park West.

Paul and his uncle believed me too literally, and after a couple of months I received a tattered letter plastered with picturesque stamps of Chilean villages, Andean summits, and guanacos. This note from Chile Chico was an anxious expression of anticipation of our imminent arrival. But it was impossible to raise the money for an expedition in such a short time, and the window of good weather had already passed—the austral summer in the Patagonian Andes ends by March, when the rainy autumn heralds a bitter winter. It was not until January 3, 1986, that I could leave New York with a small crew to make the twenty-four-hour, four-part flight to Chile Chico.

Like its sister continent in the Northern Hemisphere, South America is rich in fossils and has attracted a long line of fossil hunters. Not the least of these was Charles Darwin. Between 1832 and 1834 Darwin cruised on the *Beagle*, making various stops along the Patagonian Argentine coast, collecting the fossil bones of giant sloths and armadillos. The relevance of the place to his simmering theory of evolution did not escape his notice. In speaking of the fossils of Patagonia, Darwin wrote that he was collecting "a perfect catacomb for monsters of extinct races."

After the *Beagle* threaded through the stormy channel at the southern tip of South America that was to take its name, the ship limped its way up the rugged, ocean-chewed Chilean coast. In July 1834 it cast anchor at the seaport of Valparaíso, Valley of Paradise, west of Santiago. From this point Darwin could explore, on the back of a mule, the high mountain passes of the Andes. Here he found evidence of fossil marine creatures, as well as petrified pine trees in marine rocks, which indicated that the land was once submerged under the ocean and then raised 7,000 feet. The collection of fossils he brought back to the *Beagle* did not, however, include whales or other vertebrates. In fact, even in the 1980s, rich pockets of fossil vertebrates were virtually unknown in the Andes of Chile. In contrast, over much of the twentieth century dramatic finds of fossil vertebrates were being made elsewhere on the continent by local scientists like the great Ameghino brothers and foreign paleontologists like William Diller Matthew and George Gaylord Simpson.

One of the most colorful, persistent, and successful of these early explorers was John Bell Hatcher, who wrote a vivid memoir of his adventures later reissued under the title *Bone Hunters in Patagonia.* Hatcher was never certified as a distinguished professor at Princeton or any of the other institutions where he worked, but he was a bone hunter par excellence, and from March 1896 to September 1899 Hatcher led the Princeton University Expeditions to Patagonia. The official designation of the venture as an "expedition" was rather an overstatement. Hatcher was quick to remind his readers that during their cumulative duration the "expeditions" never included more than himself and an assistant. Despite its small scale, his paleontological campaign was extraordinarily ambitious. The duo roamed through the canyonlands and river valleys of southern Argentina during the broiling summer months of December, January, and February and endured three brutal winters on the windswept Patagonian plains. Hatcher even ventured westward toward the Andes, finding dinosaur beds and mammalian fossils in the vicinity of the isolated Mayer Glacier.

Hatcher was wildly successful in extracting a huge sample of vertebrate fossils from the remote parts of South America. But the explorer himself added a personal dimension to this venture. He was one tough character. A mishap in the wilds of western Patagonia illustrates the point. In November 1896 Hatcher decided to take an audacious 225-mile ride alone from his camp in Argentina at the mouth of the Coig River to Punta Arenas (Sandy Point), on the northern shore of the Straits of Magellan. His purpose was to supervise the shipment of precious fossils from Punta Arenas to New York. He was unmoved by the admonitions of local ranchers; in North America on a few occasions he had covered nearly a thousand miles alone and on horseback. Within a few days' ride on this Patagonian journey, however, Hatcher had a bizarre accident that nearly killed him. As he was disentangling the hoof of his horse from a loose rein, the horse suddenly panicked and lurched its head up and down, driving the broken shank of the bit into Hatcher's skull. The wound was deep and extensive, and Hatcher could not stop the bleeding even after an hour. He remounted and traveled on, but the bleeding persisted until his shirt was entirely soaked. Half-dazed and in extreme pain, he dismounted, picketed his horse with a long bone from a guanaco skeleton, and spent a nearly sleepless night while the blood from his wound slowly coagulated.

The next morning brought hope of comfort and recuperation, but after a few hours' ride he received only inhospitable treatment at the bleak settlement of Ooshii Aike. Exasperated, Hatcher forced his way into a cabin, where he cleaned his wound and consumed some food and hot coffee. As he continued his journey over the next few days he suffered from violent fever and chills, and his head wound continued to ulcerate. In critical condition, he at last reached Sandy Point only to find the medical care there less than satisfactory:

> There were two physicians in Sandy Point, one of whom had been recommended to me. I confess that the impression made upon me by this gentleman at first sight was not a favorable one, and when a moment later he suggested bleeding as the initial treatment, I was not long in deciding that I would be my

own physician and surgeon, well knowing that since the first night on the pampa after my accident I had been in no way suffering from an excess of blood. From an apothecary I procured some quinine, carbolized vaseline, absorbent cotton, bandages and a few other simple remedies and returned to my hotel.

This application of outback medicine had only mixed results. Through subsequent weeks, after returning to his camp and heading out to prospect, Hatcher suffered greatly from the infected wound. By Christmas Day it had reopened and the infection had spread over his head, face, and neck; his eyes were entirely shut by swelling. The infection gradually subsided around New Year's Day, but not before Hatcher had lost all his hair. During his raging fever Hatcher in delirium vividly experienced an expedition to Greenland, a place he had never visited. As he recounts it:

> The personnel, the party, the name of the ship and her officers, as well as our landing place and the site of our encampment in Greenland were all, not only distinctly and vividly before my imagination, but so deeply and thoroughly fixed in my mind, that it was many months after my recovery before I was quite able to convince myself that I had really never been to Greenland.

Hatcher's rugged experiences, as well as the hardy ventures of other North Americans like George Simpson and many Argentine explorers like the Ameghino brothers, opened up new territory for fossils. Likewise, discoveries made in Brazil, Colombia, and Bolivia enriched the picture of South America's grand evolutionary history. Two hundred and fifty million years ago, this continent was closely sutured to North America, Africa, and Antarctica in the great supercontinent of Pangea, and its fossils were appreciably like those of the neighboring land areas. Even after the breakup of Pangea around 200 million years ago, South America remained part of the southern Gondwanaland complex that included Africa, India, Antarctica, and Australia. Conversely, its connections with the Northern Hemisphere were intermittent and its dinosaurs and mammals evolved along different pathways from those of North America and Asia. In the latter, great Jurassic monsters like the sauropods dwindled during the Cretaceous, while duck-billed hadrosaurs and horned ceratopsians surged in numbers and diversity. In South America, sauropods persisted as giant denizens of the Cretaceous landscape, with much more modest incursions of hadrosaurs and ceratopsians. And as for mammals, Cretaceous South America showed a combination of forms similar to those of North America, like the multituberculates, along with some bizarre and enigmatic forms that were unique to the south, animals that have been given appropriate names like *Gondwanotherium*.

By the Late Cretaceous, South America had become a gargantuan island. Africa had long been separated by the opening of the Atlantic Ocean; connection with North America had been severed, and the spindly arc of land that joined South America and Antarctica had become a string of islands. There in the southern half of the world, floating free, South America became its own evolutionary experiment, an experiment that fascinated Darwin and many others. South America had a fauna that evolved over 60 million years in what George Simpson called "splendid isolation." While North America was home to a flourish of placental mammals, South America instead harbored a great radiation of marsupials, the fossil relatives of opossums and pouched kangaroos, which diversified into mammals of all shapes and habits. Large, carnivorous forms included saber-toothed marsupial "cats," with absurdly long scimitar canines that even exceeded those of the "true" saber-toothed cats of La Brea. In addition, South America was populated with the earliest edentates, the strange group that includes the sloths, anteaters, and armadillos.

There was a third curious radiation of mammals in South America. These were herbivores, hoofed animals that looked like dead ringers for horses, tapirs, camels, and antelope in the north. Yet these forms, collectively called notoungulates, were not at all clearly related to the northern taxa. Instead they showed telltale features of the teeth and skull that suggested they had arisen from some mysterious ancestors completely separate from the northern herbivores. Over tens of millions of years these notoungulates evolved

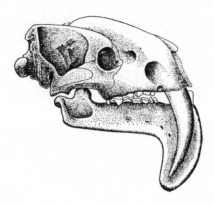

The skull of *Thylacosmilus*, a marsupial "cat" from the Tertiary of South America

in isolation, mimicking the northern ungulates in body form, limb proportions, and presumably adaptations. Some of them had stretched-out necks like camels; others had elongated snouts like ancient elephantine mastodons or mammoths. But these animals were neither camels nor elephants; they were startling examples of the power of convergence in evolution.

About 25 million years into the history of their isolation, the mammals of South America were joined by two additional groups, the rodents and the primates. We are not sure how they got there. By this time, in the late Eocene epoch, about 40 million years ago, South America had long been an island continent. There is considerable debate over whether both the South American primates, represented by the New World monkeys, and the rodents, represented by a diverse group that includes porcupines and guinea pigs, floated ashore on pieces of wood or entanglements of plants from either North America or Africa. Anatomical comparisons and considerations of evolutionary relationships favor Africa as the home port of these creatures, but resistance to this notion remains. Whatever their source, rodents and primates soon became highly diverse and successful inhabitants of forests, savannas, mountain valleys, and steppes in the tropical and higher latitudes of South America.

Rodents and primates persist today in South America, as do opossums, anteaters, armadillos, and tree sloths. But a lot of animals are missing. It would be nice to catch a distant glimpse of a saber-toothed marsupial cat or the long-snouted *Macrauchenia*, but these

The ancient mammals of South America, before the Great American Interchange

are gone. So are the vast array of ground sloths and giant armadillolike glyptodonts. Darwin himself pondered the demise of
the South American forms and, by contrast, the richly populated
savannas of Africa, across the Atlantic Ocean, which survived to the
present.

Why this dramatic downfall of the South American mammals?
Most scientists trace the beginning of the end to the very event
that brought South America back in contact with the world. About
three million years ago, near the end of the Pliocene epoch, the sea
level fell and brought a string of islands together along the spine of
Central America. These islands became the present-day isthmus of
Panama and formed a bridge that catalyzed phase two of the evolutionary experiment at the southern end of the world.

This event, frequently and euphemistically called the Great
American Interchange, was actually more of a one-way massacre.
From the north, a great tide of carnivorans, rodents, and hoofed
herbivores moved south into the reconnected continent. This invasion was accompanied by extinction of many of the resident forms
in South America, especially marsupials and notoungulates. Some
of these South American residents were diminishing in diversity
before the great Pliocene invasion, but it is clear that completion of
the geological equivalent of the Pan-American Highway allowed a
massive immigration that even many of the surviving South American lineages could not withstand. Most likely the residents were
simply outcompeted by the invaders for the finite resources in food
and habitat that typify any environment.

Despite a rough balance in the exchange of taxa between the
two continents, the imbalance in success is stark. About twenty-
three northern genera moved south. Half of the mammalian genera
living today in South America are descendants of these North
American invaders. Cricetid rodents, the group that includes the
familiar variations of field mice and wood rats, represent much of
this sweeping invasion. Their success is complemented by diverse
canids, including foxes, coyotes, and wolves; the felids, like the
mountain lion and the jaguar; and the cervids, or deer group. Conversely, about sixteen southern genera migrated north, of which
only five gained any foothold in North America that allowed their

survival to the present day—the nine-banded armadillo, opossum, porcupine, and two other kinds of rodents. This can hardly be called a massive invasion of the north by the south. Of these emigrants to the north, porcupines and the other rodents now range broadly. Opossums are extremely widespread today, patrolling with rats and raccoons the alleyways and garbage cans of North America. Armadillos are abundant in Texas but peter out because of the seasonal weather along a range limit north of the state. Their giant glyptodont relatives so common in La Brea are of course gone, as are their more distant relatives the giant ground sloths. The bizarre, two-toed tree sloth literally hangs on only in Central America and southern Mexico.

Why such an uneven outcome in an exchange of animals between two continents? The problem has fascinated paleontologists through the years, and it has elicited some interesting if not altogether substantiated explanations. Some claimed that marsupials and notoungulates were simply outcompeted because of their more primitive biology and inferior behavioral adaptability. This is a howling circularity—are marsupials extinct because they are inferior or inferior because they are extinct? It also prompts the question why such persistently primitive creatures such as cockroaches have endured for more than 300 million years. In one of his influential reviews, George Gaylord Simpson proposed that the northern species of placental mammals were destined to become the victors of the exchange because of their resilience to invasion. These forms had experienced invasion, migration, and exchange of mammals for 60 million years, he noted, during which North America was intermittently connected with Europe and Asia. The northern placental species thus had already been strongly selected for traumatic environmental disruptions that come with the invasion of alien species.

Simpson's notion seems plausible but goes way beyond what the fossil record can support. We do know that today, introduced alien species can wipe out resident species on islands, rivers, lakes, or even continents. In fact, the human-driven introduction of species to new areas is one of the primary causes of the loss of biological diversity around the globe. Often these injurious species are wide-ranging forms that have been exposed to a variety of other organisms and habitats as Simpson noted. They are versatile, adaptive, and lethally explosive in numbers when they get to new, virgin territory. The Nile perch, a common, rather ugly fish, was introduced to Lake Victoria in East Africa and promptly eliminated most of the 200 different species of fascinating, colorful, and diversely adapted cichlids that were formerly isolated as residents in the lake. There is a long list of other harmful and highly resilient species that have caused big problems—mongooses on Caribbean islands, zebra mussels in the Great Lakes, rabbits in Australia, and more. But there are also instances when very specialized, isolated forms, called endemics, resist the invaders, as is the case for some of the plant species on the British Isles. So a highly sheltered life and evolutionary history do not necessarily condemn a species to extinction when a hard-driving, cosmopolitan species appears on the scene.

Whatever spawned their diversity and caused their demise, the South American land vertebrates—from dinosaurs to didelphid opossums—have been an object of much fascination for paleontologists, and my colleagues and I at the American Museum enthusiastically shared that fascination. Like those invading mongooses we were generalists, not picky about our prey, interested in scarfing up any kind of dinosaurs, mammals, crocodiles, or turtles we could find in any new, rich fossil territory. We were first on the track of these fossil whales that were curiously stranded on high ground. The whales were probably of Miocene age, about 15 million years old and about 10 million years older than the Great American Interchange. We also knew that the southern Andes was not only one of the world's wildest and most scenic places but a vast unexplored pile of sediment—in some places 20,000 feet high—built from layers upon layers of rocks. Those rocks represented ages that extended from the time when South America was just the southwest bulge of Pangea, through the age of the dinosaurs, nearly up to the time of the extinction of those strange South American mammals and the invasion from the north. It was a terrific opportunity to see the world. And an air traffic snafu at a New York airport had suddenly and unexpectedly offered us that opportunity.

On January 3, 1986, we barely made the plane at Kennedy for

the flight to Miami and then boarded an eleven-hour flight for Buenos Aires. Our expedition was larger than Hatcher's but not by much. Besides myself it included my compadres John Flynn, Mark Norell, and Andy Wyss, as well as another curator at the museum, Malcolm McKenna. In the early morning, after a fitful sleep in a cramped seat, I pulled up the shade and caught brief sight of a huge, muddy river slithering like an anaconda through a land of dark green shadows. The Amazon, the greatest freshwater irrigation system on earth, made it clear that I was setting off for the outback of the fossilized earth—to the far end of an alien continent. I had an ineffable feeling, almost an apprehension—one I have had many times since—that I was heading for one of those few distant, mysterious, and dangerous realms remaining on our human-infested planet. Was I fit for the journey? This was no romp for geodes in a California desert. I had moments of anxiety over the loose threads that had been tied so quickly in order to launch this risky venture. Then the feeling drifted away, like the shadows over the green forest below. I smiled to myself. I was lucky—lucky to have seen Paul Raty's crude photographs of fossil whales and lucky for all the things that had come before that encounter.

18

. . .

A Man and His Horse

In paleontological exploration as in any uncertain enterprise, optimism can get the better of you while luck will only get you so far. I was soon to experience a freakish but nearly deadly incident that forever changed my casual attitude toward a certain form of transportation. Ironically, the accident had an eerie resonance with John Bell Hatcher's Patagonian mishap some ninety years before.

Our first few days in Chile were pleasant and uneventful. The plane flight from Buenos Aires to Santiago offers one of the great spectacles of the world. To land in Santiago, a city deeply nestled in the mountains, the jet started descending as it crossed the Andes; it seemed to graze the side of Aconcagua, at 22,841 feet the highest summit in the Western Hemisphere. The peaks were alive with a diamond rind of snow, but their slopes descended into an underworld of raw, ashen valleys where dark green blotches represented clusters of enduring vegetation.

As we stepped out of the plane in Santiago, the dry heat massaged our bodies, cramped and stiff from the long flight. The atmosphere—with its aridity, its faint whiff of a distant ocean, and even its smog—gave me L.A. nostalgia. We easily negotiated customs, although we had been warned about some of the scrutiny that might be expected from the security police. Chile was then governed by the oppressive Pinochet regime. Grabbing nine heavy mountaineering rucksacks, we plowed through the crowds looking for our welcoming party. The throng at the airport was hopelessly large and heterogeneous, but fortunately our hosts had no trouble recognizing us. The paleontologists Patricia Salinas and Daniel Frazinetti—the latter was going to accompany us to Chile Chico and beyond—greeted us warmly. Patricia and Daniel were specialists in invertebrate marine fossils. Except for a fossil fish expert, Gloria Arratia, who worked at the University of Kansas, there were no vertebrate specialists in Chile. And, indeed, there were no fossil mammal or dinosaur vertebrate localities of any reputation in the entire country. The two paleontologists were accompanied by an alert and extremely efficient tour guide, Rosa, who arranged our same-day flights south to Chile Chico and got us a good black-market rate for Chilean currency (192 pesos to the dollar for $4,000) while she either towed or cradled her beautiful little daughter.

What followed was a magnificent 1,100-mile flight down the spine of the Andes. We first stopped in Puerto Montt, gently alighting on a runway guarded by the augured hulk of a DC-3. Our plane rested only long enough to load fresh shrimp, crab, and oysters for the afternoon snack. It was a fitting repast for a further mystical glide over perfect, snowcapped volcanic cones, a trio of Mount Fujis, succeeded by mountains cloaked in mist, dark forests, and fjords bordered by glaciers that seemed to flow below us in scallop-shaped waves. After 400 miles, the earth below became more barren and rough-edged, an extension of the windswept Argentine plains to the east. The plane took a nosedive toward the airstrip at Balmeceda on a westerly bearing, straight into a wind that seemed strong enough to shred a windsock.

The wind on the tarmac was indeed formidable. This was the

land of the "roaring forties," the most notorious and persistent winds on earth. Silvery tufts of grass shivered in this gale, sparkling like lightning in a sun that seemed to sail with the clouds over the plain. People were impressively oblivious of the wind; they went about their business offloading heavy sacks of flour or plaster and cumbersome boxes containing television sets, mummified with miles of fiber tape. I tried to imagine a normal existence with a gale blowing in my eyes every waking minute.

We planned to join throngs of people in the rustic terminal and board some buses for the provincial capital of Coihaique, where we would spend the night and return to the airstrip the next day for the one-hour flight to Chile Chico. Daniel, however, had a good plan to bypass this layover and booked us on Air Don Carlos for a flight late that afternoon. Some apprehension had been expressed about this carrier back in the States. We were told by American airlines that they could not book us through Air Don Carlos on the last leg of the journey because the company had not been approved for safety. In any case we boarded a twin-engine Cessna loaded to the gills with our six expedition members, our heavy gear, the pilot, and two young girls. The pilot, a nervous but pleasant chap, anchored the bulkier of us into the plane's center of gravity and secured everything with a foreboding complexity of straps and ropes. As we took off, we could feel the plane stall and seemingly float backward in the wind. Malcolm, an amateur pilot with many years' experience, was white-knuckled as he observed, "The stall light's been on for the entire takeoff!"

The Cessna barely crested the mountains south of Balmeceda and we sailed over the glacier-fed, blue-green ellipse of Lake (Lago) General Carrera (which receives a different and more familiar name, Lago Buenos Aires, on Argentine maps). At this point hundred-mile-an-hour gusts nearly blew us into Argentine territory. But the wild flight did not distract us from one of the most stupendous geological landscapes any of us had ever viewed from aloft. The lake, which straddles the Chile-Argentina border, is ensconced within the arms of huge peaks hung with silver glaciers. To the south are austere black needles of eroded volcanoes. And all around are layers of rocks—pinks, browns, gray-green—hundreds of square miles of rock, perhaps many of them with fossils and virtually none of them ever walked over, let alone prospected. Several lifetimes of fieldwork stretched out below us.

The crowd at the Chile Chico airport, near the Argentine border, consisted of an elderly lady, presumably the grandmother of the young female passengers, and an orderly whose ambulance carried a patient bound for the return flight to Balmeceda. From the plane a cab had been radioed for, to take us to town, where we were greeted by a very jolly mayor, Carlos de Smet de Olbeck de Halleux. Inside the mayor's house, sounds of children were interspersed with the stern voice of an old woman barking orders in French. I gave Carlos a bottle of Glenfiddich scotch and we tried to carry on a bilingual conversation, aided immeasurably by his wife Beatrice, who spoke English with brilliant, almost poetic clarity. The mayor was excited to start planning for our assault on the isolated summits with their beached fossil whales, but he could see that we were now hardly conscious after the twenty-four-hour flying marathon from New York. He clapped his hands and ordered our cab to convey us to the Nacional, the only hotel in town.

The next morning at Carlos's house we planned our campaign. Maps were opened and strewn about. Geological colors indicated rock formations that stretched for miles in every direction, and it would have been a bewildering exercise to narrow our targets for exploration, if not for the successful reconnaissance of Carlos and his nephew Paul. From this lengthy conversation, two targets had been identified. The first was Pico Sur, south of Chile Chico, the mountain where the whale bones had been recovered. We decided to take nine people, including two guides, on horseback to Pico Sur. The second target required a boat trip westward along the southern shore of Lago Carrera about fifty miles to a small outpost known as Mallén Grande. Information on fossils from this remote area was sketchy but intriguing. Carlos showed us a photo of a fossil taken by a teacher in Mallén Grande. We were at once struck by the fuzzy image of a skull of a large animal, complete with teeth. We stressed the need to get to that place, even with a small team of four or five people. Carlos promised to ensure this, and we drank to our good fortune with glasses of *piscola*, an overly sweet concoction

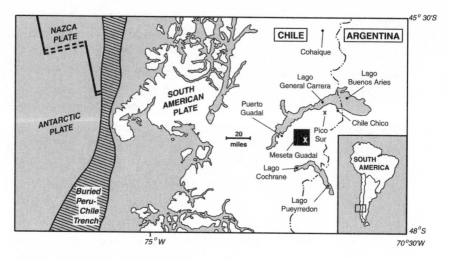

Important towns and fossil localities in the region of Chile Chico, in southern Chile. The fossil locality Pampa Castillo is marked with an X within the Meseta Guadal area

consisting of Pisco—a strong, tequilalike grape liquor—and familiar Coca-Cola.

Then came four days of waiting while horses were gathered, shoed, and supplied. With a bit of Yankee impatience, we hung out at the Hotel Nacional, dining on bland or in some cases rather repellent food, complete with insects in the meat pie. Nonetheless, Andy was exceptional in adapting to the meals. He was, after all, very content back in New York with the cuisine at The Food Shop or with his own concoction: pork and beans heated in the original can on the stove and garnished with frozen peas. The meals were the social events of the day. Much of the time we spent on our own, catching up on manuscripts or practicing Spanish in our rooms while outside the Patagonian wind made the telegraph lines shriek.

Carlos took us to the top of a grassy hill just west of town, where a kind of monument was ringed with flagpoles. Every Sunday, flags were unfurled representing the home countries of the residents— Brazil, Argentina, Belgium, Germany, Chile, France, and Italy. It was a colorful display of internationalism for the proud people of Chile Chico—little Chile—so named because all that was Chile—lakes, snowcapped mountains, trees, *estancias*, grassy pastures, fruit orchards, sun, and of course wind—was contained in the pleasant, scenic countryside that surrounded this town.

At dusk we returned to the hill with the flags, and looked down on the well-disciplined grid of gravel streets lined with shimmering poplars, groomed like the rows of trees around the Jardin des Plantes in Paris. The amber streetlights floated like fireflies over the shadow cast by the Andes. Above I had my first view of the brilliant star-studded southern sky. Orion was upside down and unfamiliar in frame, a leaner and hungrier hunter than the Orion I used to study as a boy in the winter sky over northern Wisconsin. The megastar Aldebaran in Taurus shone as brightly as Venus did in the northern sky. But there were new elements of drama in this southern vault. The Magellanic Clouds and the Southern Cross hung over the daggerlike black silhouettes of sculptured volcanoes near Pico Sur. I could not imagine a more arresting scene on earth.

The languid ennui of our first days in Chile Chico was interrupted by an excursion on January 7 to an area southwest of town, where two years before Carlos had found a rich layer of fossil oysters in a deep canyon. The oyster beds themselves were nearly vertical, a canyon wall that plunged down to a gravel-choked stream. These represented what had been a very shallow lagoon probably at the edge of a great bay that swept westward from the Atlantic millions of years ago. We also found barnacles, brachiopods or lampshells, sponges, clams, and snails, but nothing of burning interest to the vertebrate paleontologists in the crowd. Daniel Frazinetti had a particular research interest in the oysters, though, and planned to spend most of the day at the spot. The rest of us were itching to explore the mountain wonderland that loomed above us.

After a pleasant lunch of cheese, salami, crackers, and chocolate—a break from the entomological meat pies at the hotel—Carlos, Mark, and I headed up the steep mountain slopes, prospecting some sandstones just west of the oyster beds. There was little chance of good fossil sites along this trail, but the view was spectacular. We could see Pico San Nicholas, the giant eroded neck of a volcano that dominated the mountainous southern horizon. To the north we had a view of the jagged crown of Cerro Castillo, shrouded with huge glaciers. Carlos explained that the row of

Chamonix-like needles that formed the summit of Cerro Castillo had attracted many top climbers, including a couple who had recently died there. We were now in an alpine meadow, braided with small springs and creeks and filled with lady's slippers, daisies, and wild orchids. The flowerbeds reflected the sunlight like a multihued prism with a predominant splash of yellow. *"Muy bonito, eh?"* a buoyant Carlos exclaimed regularly, seeking well-deserved approval of his homeland.

As we climbed to the top of the peak above the meadow, the inevitable Patagonian wind almost swept us off the mountain. My glasses were the only casualty; they were lifted from my face and seemed to float for a moment in the Andean atmosphere before plummeting down the slope. An hour's search yielded only fragments of my spectacles. I had a replacement pair but unfortunately these were back at the hotel. The loss made for a very blurry descent down the mountain, past grottoes of pink, red, and green rocks matted with the yellow of alpine flowers, looking all the more dazzling, if undefined, in my nearsightedness. It was a glorious day, our first romp in the wilds of the Patagonian Andes. It even made tolerable that night's entrée at Hotel Nacional—some spherical lumps best translated as "tripe balls."

At last we were ready for the Pico Sur expedition. At 7:00 a.m. on January 9 we drove in an open pickup to an *estancia* that claimed a beautiful green clearing and a stand of poplars. We loaded up a couple of hyperactive packhorses, *pilcheros*, with our supplies for nine people over five days. When we mounted our steeds, our *baqueanos*, or guides, were not happy about the lugs on the soles of our mountaineering boots, which fit all too snugly in the open stirrups. A couple of sheep were slaughtered and loaded onto the packhorses, while the dogs snapped and fought over the discarded entrails. With the butchery completed, we started up a twisting trail along a huge precipice.

As the hours passed we slowly gained altitude. One thousand feet higher and the trees became stubble. Two thousand more feet and the thick grasses and stunted trees faded. Two thousand additional feet brought us to hanging gardens of alpine flowers. Looking north over that great expanse, I saw a world almost completely in shades of blue—the great lake, the glacier in shadow, the distant rock faces reflecting the snow, and the sky above. Only the base of the panorama broke with this color scheme: some 5,000 feet directly below one saw deformed, metamorphosed ridges and folds of pastel grays, greens, and purples that swirled around the edge of the lake, like those colorful flags on the hill near Chile Chico. These rock formations were set in a nest of silvery grasslands that looked like an endless Welsh moor.

The horses could not enjoy the view, however. They trudged up the steep trail like laboring locomotives, huffing and emitting a foggy breath. Our two *baqueanos* kept a careful eye on their condition. In the lead was Crus Vargas, a resplendent cowboy in loose black clothing and high boots, sporting a broad-brimmed flat-topped hat and a long, wicked knife. Much more casual, in a bulky wool sweater and a wool cap that sometimes covered his eyes, was the younger Roberto, or Bobby, or, as Carlos and company called him, "Boy." Bobby would whoop and strike out ahead of the pack, then return to smile and flirt with two comely companions, Nickette, the mayor's niece, and Vivian, a close friend of Nickette's from Argentina. Whether guide or client, friend or family, all our South American companions had one thing in common—they could ride superbly. We watched their antics with amusement and I thought how much more wild and exciting the whole affair was than loping up one of those trails in the Grand Canyon on the back of a mule.

Soon, though, even the enthusiastic *baqueanos* seemed tired. Crus ordered us to rest the horses on a sloping shelf of grass and boulders little wider than the roof of a Patagonian ranch house. As I fiddled with my camera, my horse, obstinate, tired, or disoriented, suddenly buckled underneath me. In an excited voice, Bobby ordered me to dismount. I was already making the effort: I swung my right leg over the flanks of the horse, but my left foot was trapped in the open stirrup. Unfortunately, the horse now stood upright again and started bucking, and dragged me over the clearing, which was studded with sharp-edged rocks. There was much shouting, including my own expressions of terror. The horse halted, but before I could dislodge my left boot it took off again. My head hit a rock just before the horse reared and seemed to come down on me, grazing my right leg. There was an excruciating amount of running, dragging, and shouting.

The pandemonium of this unplanned rodeo sent ripples of panic through the whole pack of horses. We were all on the verge of disaster, for if the spooked mounts peeled off the ledge, they would take the quick air route back to Chile Chico. But at last the *baqueanos*, with Mark's assistance, held my horse in check, and I squirmed out of the stirrup, my arms folded closely against my body in case the horse should rear again.

I lay on my back with my bashed head leaning against the jagged pillow of a rock. With some anxiety I saw that my right leg was encased in bloody, shredded denim. I felt nothing. Compound fracture? Oh what the hell, I felt lucky to be alive. Mark and Boy came over and felt my legs. I could indeed move both of them, a seemingly impossible outcome. Crus nodded approvingly, then muttered *"Caballo malo,"* his unfavorable assessment of my poor, sick horse. Mark wrapped my leg in a tight bandage.

There was much discussion about what to do, with some thought that maybe I should return to Chile Chico. "What's in Chile Chico?" I protested, thinking of those tripe balls at the Hotel Nacional. "Besides, the worst has to be over, we're almost there." Within an hour I was hoisted onto a trustier steed, a large white horse that also carried a sheep carcass over the back of the saddle. Sheep and human blood commingled on the back of that poor horse as we joined the rest of the train.

A scenic, exciting trip was now transformed into an endurance test, a hellish three-hour ride. The cold streaming off Pico San Nicholas penetrated my torn jeans and froze my stiff, painful legs. My face was so swollen from hitting the rock that I couldn't open my mouth. The distant blue Andes were a mirage blurred by intermittent sleet. The events around me were as hazy and dreamlike as that mountain skyline. A roving sun highlighted the golden back of a guanaco standing on a distant ridge. There were shouts from Boy, Vivian, and Nickette. Another guanaco, then two giant condors. A large ram guarding a flock of sheep challenged one of our skinny dogs, called "guitars" by the *baqueanos* because of their contorted fiddlelike bodies and slender necks and limbs. The ram's threat was a strategic error. The dog took out after the ram at blinding speed and bit through its head. More blood.

We ascended a steep ridge onto a high plateau whose surface was cut by a maze of icy creeks. I felt the cold water against my burning legs. Pico Sur, a high, snow-crowned butte, lay directly to the west. At last, at 3:00 p.m., we arrived at a small meadow below Pico Sur, our intended campsite. The wind was bitter cold at 10,000 feet, and I was wrapped in every layer of clothing I had packed. The *baqueanos* gathered sage and roasted a side of a young sheep—a *cordero*—for more than three hours over the fire. The Chilean and Argentine clan—man and woman alike—sliced into the meat with their sharp hunting knives and chomped down on big chunks. The *cordero* was smoky and tasty, especially when slathered with a bit of *aji*, or garlic sauce. By the time we were finished with our meal Crus and Boy had constructed an ingenious shelter of tarps, blankets, sheepskin, and saddles between two rocks. The dogs and humans huddled together for warmth, but eventually the cold winds drove us back to our tents for the night.

I awoke the next morning at this camp, still frigid in the midst of the austral summer, with painful legs and a swollen jaw. Nonetheless, I was feeling strong enough to limp after the others on the long ascent up another 1,500 feet to the whale site. The fossil beds, which we called the Amarillo Beds, were buffy gray to yellowish, with lots of oysters weathering out on their surface. Immediately we found some big whale vertebrae in the streambed below the main fossil lens. Malcolm got a few nice shark's teeth; I found a variety of bones in the yellow layers near some volcanic outcrops. But prospecting and excavating was annoying work in the muddy, wet sediments.

Things became less comfortable as the wind picked up and brought in an ice storm. There was a brief respite when the sun warmed us. For lunch we downed some sausage and cheese, after which Mark and I prospected upslope in high winds that nearly blew us off the ridge. When we returned to the main site, everyone had retreated to camp except Malcolm, who was off in the volcanics looking for rock samples for radiometric dating, and John and Andy, who were soundly asleep. The exhaustion of yesterday's ride, the cold, the wind, and the rather disappointing cache of isolated whale bones and other fossils made us look forward to the next phase of exploration in Mallén Grande.

My legs were stiffening in the cold and I headed down to camp. Soon I was joined by Malcolm, who brought back a choice, elongate tooth of *Isurus*, a nasty fossil shark. John and Andy straggled back to camp, speaking rather despondently about the eighty-mile-per-hour winds at the summit of Pico Sur. But the wind had died down in our camp and we settled in for some restorative *maté*, the thick, slimy green tea that is a strong stimulant and staple for the South American cowboy.

Dinner was a delicious stew of *cordero*, rice, onions, and *aji*, and people were festive despite the day's meager finds. There was card playing—a game called Truko, which required elements of deceit similar to our North American game "Bullshit." Poor Vivian was teased about the recent disastrous defeat of the Argentines by the British in the Malvinas or, as the Brits preferred to call them, the Falkland Islands.

The kidding and camaraderie among these country people belied the tensions at official levels between Argentina and Chile's Pinochet regime. The Chilean leadership resented the fact that Argentina had ended up with oil reserves in their part of Tierra del Fuego, both countries claimed overlapping sections of mineral-rich Antarctica, and there was talk of a possible war when the Antarctic treaty was due to expire in a few years. Conversation ranged from politics to paleontology, to the medicinal powers of garlic, to much-requested descriptions of the fun, sun, and surf in "So Cal." These Patagonians had a particular slant on the outside world: "Norte America" was some vague landmass in another hemisphere, but the crime-filled streets of Nueva York and the glamour of Hollywood were vividly appreciated. Our happy conversations faded in the wind and the cold that abruptly terminated another evening at Pico Sur.

Sometime after midnight I woke up with a need to urinate. To my alarm, I discovered that my legs were hardly functional; both knees were so stiff I could not bend them. I wriggled out of the tent into ghastly cold; the mountain meadow glistened under a night sky dusted with the brightest stars I had ever seen but at the moment could hardly enjoy. To relieve myself I propped my inflexible body against a rock. It was an ordeal to crawl back into the tent. As I returned to that abyssal sleep that only sheer exhaustion can inspire, I pondered my prospects as a field paleontologist tomorrow and beyond.

The answer to that question the next morning was clear for the near term. There was no way I could join the others, so I lay about camp like the wounded on a cold alpine battlefield. Boy heated a Pisco bottle, which Nickette secured against my more painful knee with one of my T-shirts. The cold and the wind were unrelieved, and the water in my drinking bottle had frozen. I lay shivering near the fire, contemplating the possible impact of my injury and trying to enjoy the grand clouds frothing off the peaks above.

Carlos had led our team to a place with petrified wood, where he claimed there were also dinosaur bones. These did not materialize, however; the "dinosaur bones" were simply long, limblike pieces of wood. As we huddled under a pile of sheepskins in the late afternoon, a gale mixed with snow came howling into our mountain meadow. A horse sideswiped by the wind was actually blown over. Pico Sur was becoming increasingly unattractive.

The next morning was only the third of our five planned days at Pico Sur, but the elements had vanquished us. The cold wind picked up and now was blowing snow. It was indeed a rather desperate predicament, and Carlos advised a hasty retreat. My replacement horse was docked against a big rock, and I was lifted into the saddle with my unbendable legs. It seemed ludicrous to contemplate that my only salvation lay in another eight-hour ride. The horses themselves were not happy about the journey. They bucked and bit each other as Boy went around carefully checking all the cinches. As John tentatively mounted his horse, it bucked him so violently that he made a graceful somersault over the horse's head.

This comedy of confusion in our alpine meadow, the rout of an expedition, was interrupted by a sight of mountain gods. Boy and Crus started yelling and pointing up to a ridge enshrined in clouds several hundred feet above camp. As the clouds broke I could see a *baquero*, a cowboy, on a Pegasus, a stallion that seemed to fly from one basalt promontory to the next. The *baquero* whooped and waved down to us, obviously enjoying his theatrics. Each time he took a death-defying leap our guides would sing out in approval. Like Hatcher's vision of Greenland, I could not ascertain whether this valiant rider on his magically airborne horse, and the Olympian cloud-covered summit, were real or merely a hallucination, but others testify to their reality.

We returned to more serious, secular matters. Crus decided that the bad weather meant almost certain disaster for a descent to Chile Chico the way we had come. Instead he intended to lead us due east on a radical descent through the great volcanic spires to the foothills and the pampas near the Argentine border. But while this gave us protection from the bad weather, it exposed us to a far steeper and more treacherous route without a clear trail. After shivering in the snowstorm for a half-hour while Carlos and the guides debated our options for escape, we at last were ready to go. We started east on an uncharted route; Crus rode attentively back and forth, reminding us to keep a very tight grip on the reins. My legs were throbbing in pain.

As we were riding down a steep ravine, one of our *pilcheros* suddenly erupted in panic, convulsing its body and kicking its legs. Pots, pans, and unopened cans flew in all directions, looking like a metallic cloud of ticker tape. As the *baqueanos* repaired the situation, my horse drifted about nervously, trying to find more level ground. Our mounts also indulged in a crafty trick: they deliberately bloated themselves when the cinches were tightened. Then, when the horses exhaled, the straps became looser and less oppressive, but caused the saddles to slip sideways on their flanks. This deceit was particularly a problem for big John, who for a few heart-stopping moments seemed to be "mounted" under, rather than atop, his horse.

At last we resumed our descent. At least the storm was now above us. Insulated for a moment from the shrieking wind, I could hear the rush of a cascade far below. As the ravine became even steeper, the horses planted their front feet into the soggy silt and slid as if on skis. I felt a bit nauseated from the weightlessness and imbalance that came with this sudden plunge. Finally, we reached the torrent at the bottom of the canyon and were ordered to dismount—a ludicrously impossible maneuver for me. Because the others were preoccupied with their own problems, I could see that help in dismounting was not immediately forthcoming. So I simply held my arms out in front of me and rolled out of the saddle and, unintentionally, onto a thorn bush. For a moment I lay in the thorns, with bleeding hands and immobile legs, feeling thoroughly miserable and expendable.

The riderless horses were beaten into fording the river, and we soon followed, requiring no further incentive. I somehow managed, with the help of Andy and Mark, to get from one boulder to another that marked a safe route among rapids and whirlpools. I felt some relief in the crossing as the sun warmed my back and I bit into a piece of chocolate. "Crus says this is the last big river," Mark reported. But Crus was looking at the bad weather that tailgated our retreat, and he was in a hurry. My comrades hoisted me up the flanks of my restive horse, and we pushed on.

A descent into a wild ravine requires a correspondingly challenging ascent up the other side. We rode up the steepest escarpment yet, and soon we were much higher than our position had been on the opposite ridge. Crus stopped for a moment of indecisiveness and craned his neck as if he were trying to catch a glimpse of the terrain far above us and well out of view. Andy looked at me and shrugged. Where were we going? The horses protested over the precarious stop and started leaning down the path of least resistance, back down toward the river. I could tighten the reins but my useless legs could not nudge the flanks of my horse. As the mount suddenly cut back for a river view, I almost peeled off into the abyss.

We managed to turn the caravan back on course, and we were soon at the top of the ridge. A faint trail wound through a plateau leading to the foot of the great black volcanic spires of Pico San Nicholas. We were riding at the edge of the earth. Even my agony and exhaustion could not detract from my awe at that incredible topography. Our horses hugged the narrow trail at the base of a monolith of frozen lava that looked like the superstructure of a battleship. They moved forward gingerly as small chips of lava shot out from under their hooves. When the wind died, on rare occasion, I could hear the glassy chink of fragile rock and the horses snorting. I could also feel the subsonic thumping of my heart under a layer of sweaters and a nylon wind shell. My legs now felt like torches. It was a nightmare on horseback. All the while, we navigated this extraordinarily arresting landscape, which under different conditions would be hard to ever leave.

After a couple of hours of this very delicate traversal of the crumbling volcanic scree, we arrived at a long slope of tawny grass. Out of the Andes and into the eastern foothills at last! The horses naturally just wanted to rest and eat in this paradise, but the "guitars" ran back and forth, snapping furiously at the horses and encouraging their progress. Then our descent was stalled because Crus grew uncertain of the route. He rode tentatively forward but soon picked up speed as the terrain became more familiar. The way was still difficult, with steep hills and deep gullies. Even Crus had to contend with the exhaustion and obstinacy of his mount, which gave a tremendous buck in protest to yet another steep slope. Andy and I watched in amazement as Crus, momentarily airborne, held on to the reins, landed back in his saddle, yanked the horse to the left, and forced it down the hill.

Soon we seemed to be nearing some kind of destination. The mounts were compelled to make an agitated two-foot leap over an old range fence. The "guitars," however, were relaxed and thoroughly liberated; they took off after a rabbit that seemed to be leaping toward the Argentine plain on the far horizon. More fences followed, then some mudflats, another hill, another fence, a deserted house, a corral, and a sheep shed. We halted our exhausted caravan next to the shed in the shade of some huge poplar trees. The air was warm and salubrious, a warmth that I had all but forgotten over the last several days. Again I intentionally took a clumsy fall off the horse and landed in a hay pile, without any intention of righting myself, let alone managing to walk. I had not used those legs for nearly two days. But the worst commute home from work I had ever taken was over.

19

• • •

Pampa Castillo

The *estancia* we had reached on horseback, which marked our salvation from the raging winds and treacherous trails of Pico Sur, had an otherworldly serenity. It hardly seemed to be part of Patagonia at all. The majestic spires of Pico Sur, now in the far distance, kept the cold wind away, and the sun was warm, almost hot, as we peeled off our sweat-stained outerwear. Nearby in a fenced patch of grass were odd austral flightless birds, rheas, that Andy and Mark delighted in chasing. The others helped me migrate to a nearby cabin where an old man with hands that looked like basalt from the mountains above boiled some maté and fried *tortas*, lumps of bread cooked in sheep fat. As the breeze bent the poplars, I thought of those summer days in Crazy Mountain Field, Montana. Soon a pickup arrived and I was helped into the back for a bouncy ride to the Hotel Nacional.

My arousal from a deep sleep on the morning of Monday, January 13, was laced with apprehension. How functional were my limbs? As I flexed my legs to get out of bed, I felt the sharp jab of pain in my knees. But I managed to stand, and to my surprise I almost felt better in that position. The warmth of Chile Chico seemed to bring some healing. Walking was painful and laborious but possible. This improvement eased the dilemma we discussed at breakfast. According to our schedule, Malcolm was to leave the next day to join his wife in Puerto Montt, and if I went with him I would have an escort part of the way back for treatment and recuperation. But I decided to stay and see how my legs held out, if the others did not think my presence an overwhelming burden. Everyone was confident and supportive, and I limped through town on brief shopping ventures in preparation of the next phase of the expedition.

As we stopped at the local bodega, the *fruitaría*, the blacksmith (for better and more pricey horseshoes), and the telegraph office, we were greeted with warm smiles and friendly chitchat. Word of the aborted but epic Pico Sur expedition had already gotten around, and we were local heroes. Friends would encounter us in the street and ask how my injuries were doing or offer to help me walk up a step into a store, a painful and difficult maneuver. We were held in an esteem that contrasted with the reputation of the other foreigners in Chile Chico. These were ten rather pathetically prepared young men and women on one of Prince Charles's "Raleigh Expeditions," a program resembling in spirit those of the U.S. Peace Corps whereby groups of wandering youths contributed labor and teaching to "underdeveloped" communities in distant lands. Carlos had skeptically accepted the help of this platoon, an assemblage of representatives from the United Kingdom, Australia, the United States, and New Zealand, but soon realized they were rather lazy and inexperienced. When he assigned them the ignominious task of cleaning winter stables of mountains of manure, the Raleigh team protested that such work was beneath their station. Carlos promptly offered them the option of "improving" conditions in another town.

The Raleigh group hung out in a cavernous old adobe house decorated with dirty clothes and backpacks in khaki and camouflage. The place looked like the bunker of a defeated army. The

Raleighs complained that the mayor and other village dignitaries had ignored them while showering hospitality on some Yankee paleontologists. When we revealed ourselves as the beneficiaries of this flagrant favoritism, some of the Raleighs completely changed their tune, begging us to take them along, willing to desert their own troops in the process. One got the impression that deserters were not shot on sight or, for that matter, even noticed. They seemed a band without a leader or purpose. There were some reasonably nice people in this disoriented clan, but unfortunately the conversation was usually monopolized by an obnoxious girl from Leicester whom we nicknamed Gordo.

"Where's your leader?" I asked Gordo.

"Oh, him, he's trying to gather his wits out back." She motioned to a dark passageway punched out of a wall.

As we entered the dim room we saw a bulky man in military camouflage and high boots at a ham radio.

"Halloo, you the Yanks?" He looked up at us with his huge, dead pupils as if he were looking at nothing at all. His drawl betrayed the influence of barbiturates. I thought of Mr. Kurtz in Conrad's *Heart of Darkness* or Francis Ford Coppola's derivative film, *Apocalypse Now.* The platoon leader returned to his radio. "Excuse me, chaps, got to call Coihaique, our central camp."

There was some buzz and static and voices as faint as distant calls against a Patagonian wind.

"Roger. Yeah, well . . . I'm working on the mayor . . . almost got kicked out of town today . . . may have to go back to Coihaique."

There was a crackle that sounded like an exclamation and a rebuff. Apparently Mr. Kurtz and his band were not even welcomed by one of their own brigades.

The stoned commander cursed and shut off the radio. "We're having a hell of a time here. Can you help us with the mayor? How about us joining up with you? We were supposed to work on a bridge, but no one paid any attention to us. Then the mayor put us on manure detail. That means the whole town, whole place is the shit hole."

We were not inclined to any collaboration with the group, but I said that I would talk to the mayor.

In a gesture of cooperation we returned to the Raleigh group after dinner, bringing them a large liter bottle of *tinto vino* and some food. The situation had worsened. Helen, one of the team members, had lost her money pouch and the significant equivalent of $250. The team became very agitated, especially Gordo, and reported the event to the police as a theft. Gordo looked at us suspiciously, asking if we'd seen the pouch during our earlier visit. Ignoring her, we briefly chatted with the more affable and alert members of the crew and drifted out into a balmy evening and the omnipresent wind.

I savored the idea of a few days of recuperation, unlike our waiting period before the Pico Sur foray, but our tight schedule would not allow this. The next day, January 14, we would start on our voyage westward on the lake to Mallén Grande. The weather was providentially calm, meaning smooth sailing on waters that were often violently stormy. After saying goodbye to Malcolm we trucked with our gear to the dock to await the ferry, aptly christened the *Pilchero*, after the weather-beaten, uncooperative old pack-horses of the Andes. The weekly arrival of this rusty vessel, always late, was the big event in town. The whole population crowded the docks as some big Mercedes trucks were offloaded. Then a swarm of pedestrians in numbers that must have challenged the cavernous capacity of the ship walked down the front ramp of the ferry. As soon as the throngs were ashore, they were replaced by the crowds on the dock as well as trucks, a flock of sheep, and scores of feisty horses.

Carlos escorted us to the *Pilchero* and introduced us formally to the distinguished captain and a small detachment of soldiers. Out of the corner of my eye I saw a few of the forlorn members of the Raleigh group taking in this ceremony from the dock, doubtless envious of our royal treatment. This reminded me of my promise and I mentioned their plight to Carlos. The usually cheerful mayor grew stern and shook his head. *"Malos hombres,"* he said. He then quickly brightened and slapped me on the back, giving me a fond farewell and a *"Buena suerte!"*

The *Pilchero* chugged slowly away from the dock as we waved back to Carlos, Beatrice, Nickette, and the rest of the de Smet clan.

On an unusually windless day we floated on a lake of glass that reflected every boulder, crevice, and ice field of the towering peaks
above. Our first stop was Fachinal, a minute village of no more
than a quartet of houses, where a jeep was miraculously stuffed into
the car bay of the *Pilchero*. Here the road trending west from Chile
Chico abruptly ended, its connection blocked by the stupendous
mountain rampart of Cerro Fachinal. The only land route connecting Fachinal to the western village of Mallén Grande was a
narrow horse trail like the one we had traveled near Pico Sur. Crus
had identified this as a perilous route; several riders and horses had
plummeted to their deaths in the lake a thousand feet below.

Before we continued due west to Mallén, we traversed the lake
to its northern shore and a company mining town called Puerto
Cristal. The town had the forlorn look of those deserted ghost
towns I had once explored with Mick and the boys in the Arizona
desert. Here the mountains seemed even higher and more forbidding than any place we had yet visited. Their elegant, fluted
summits of ice contrasted with the squalor below, a ramshackle
congregation of huts, rusty junk piles, and muddy streets plastered
with horse dung. A series of switchback trails that crisscrossed in a
bewildering lattice extended upward on a steep, barren hill and terminated in ugly craters and fissures. From these open wounds talus
and trash from a lead and zinc mining operation spewed out over
the hillside. The place was an ecological disaster in the middle of an
enchanting wilderness. People lacking the energy and cheeriness of
the residents of Chile Chico stood passively on the dock as some
dirty burlap sacks were unloaded. Many of the men on the docks,
probably miners, were gaunt, crippled, and disfigured, as if their
burdens had included the entire weight of the Cordillera. I wondered if Puerto Cristal also served as a gulag for political prisoners,
enemies of the Pinochet regime.

A light rain fell as men on the ship sat in truck cabs and passed
maté cups, and groups of women and children wrapped in llama
blankets crowded under the trucks. The *Pilchero* did its offloading
for an interminable hour, then abruptly embarked, nearly deserting
the few boarders on the dock lucky enough to leave this depressing
port.

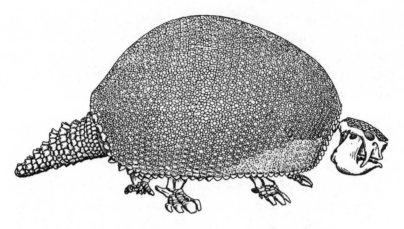

A glyptodont, about nine feet long, a Cenozoic mammal of South America

Our slow trip back to Mallén Grande and the southern shore of the lake brought back the sun and the warmth. Near evening, we reached our destination, a village at the apex of a serene inlet, hugged by some gracious, rounded mountains covered with *Nothofagus* (southern beech) and thick, rusty scrub. Farther removed from the high peaks with their glacial runoff, the still water of the bay was transformed from milky blue-green to sapphire. The bay was guarded by small islands of granite and schist studded with the dark beech trees. After the destitution of Puerto Cristal, Mallén Grande was an Aegean paradise.

The bottom of the *Pilchero* scraped the pebble beach and we promptly dragged our heavy packs ashore. There we were met by a small dark-haired man in his early thirties with a scraggly beard and mustache not unlike our own. *"Hola!"* greeted Professor Robinson Cárdenas, the local schoolteacher and the discoverer of the fossils of Mallén. Next to Cárdenas stood our intended *baqueano*, a short, very thin man in black, wearing spurs with the longest spikes imaginable. The *baqueano*, introduced to us as Pedro Soto, gave us a look of haughty reserve, as if he were measuring our fitness for the wilds. I noticed him stare at my limp as I struggled with one of my packs. As Soto lit a cigarette he reminded me of an odd hybrid, the face of the French movie star Jean-Paul Belmondo disadvantageously crossed with that of Bob Denver, star of *Gilligan's Island*. He seemed a paradox, a stylish austral cowboy who failed to project any of the true grit of a Crus Vargas. Pedro's secret nickname was henceforth Gilligan, a name so appropriate that even our Chilean comrade Daniel Frazinetti laughed in recognition.

Our transport from the bay, a hefty semitruck, was first loaded with an enormous diesel motor. I could barely ascend the tire and the wheel well with my wretched legs, and Andy grabbed my arm to pull me over the top of the truck bed. As we rumbled along the splendid Mediterranean coastline above the blue cove, I reflected somberly on my lack of fitness. Was I a team leader or merely a burden? I leaned on some saddles and looked to the south, where the naked summits rose above a dark tree line of striking regularity. Somewhere in those distant peaks were fossils of strange beasts that Robinson Cárdenas had discovered. Would we have similar good fortune?

We got off at the center of Mallén Grande: a sheep-shearing shed, a blue-and-white adobe schoolhouse fronted with a naked flagpole, two small rectangular wooden houses, and a more modern house with a bright orange metal roof and an antennae—the radio station, and our only contact with Chile Chico and the world beyond. The schoolhouse turned out to be the nucleus of Mallén—acting as courthouse, community shelter, and guest quarters for a group of scruffy paleontologists. Here Robinson began his lecture on his fossil findings. At first we were bitterly disappointed; the skull we had seen in the photograph at Carlos's house had been stolen, claimed Cárdenas. But he did have a pile of other enticing bones, a long black canine that could have belonged to a huge lumbering rhinolike toxodont, several finger bones, and something that looked like a chunk of a turtle shell except for a curious series of starlike ornaments on the hexagonal plate.

"That's a glyptodont," I said to Cárdenas, who looked very puzzled. I drew an armadillolike, spike-tailed glyptodont on the chalkboard. We were very excited. These fossils were typical of Miocene series found in 15- to 20-million-year-old beds in Argentina, but they were unknown in the Andes this far west. We anticipated with enthusiasm that there was more at the secret site than this scrappy pile of bones.

But where was the site? The answer was more elusive than we anticipated. Cárdenas did not quite remember the way to it. In the difficult conference that ensued he was joined by Eduardo Hermasillo, a somber, sluggish fellow who was to serve as our other *baqueano*. We pulled out the topographic maps, but they were hardly of value. Eduardo drifted his finger indecisively back and forth across the contours south of the lake, scratching his head. At last he recognized the course of a river, but he claimed the map gave it the wrong name. He knew it as Río Hernández.

"Who was Hernández, a famous general?" I asked, thinking of all the other place-names in the region that had military inspiration.

"No, just a man; he drowned with his horse in the river," was Eduardo's reply in Spanish. We wondered if there would someday be a Río Novacek or a Río Norell, Flynn, or Wyss in this impenetrable wilderness.

We continued inquiring about the route to the fossil beds. Their location, Cárdenas at last said, was not at all clear. No matter, a man named Ojuay who lives in a cabin near the edge of the forest would guide us to the area. Eduardo noted that the trip must be made on horseback and the rivers must be crossed in the early morning, as they run dangerously high in the afternoon. "How far is it?" we asked. The professor thought seven or eight hours, Eduardo nine or ten. At this moment Pedro "Gilligan" Soto came in and helped himself to *maté* and *amasado*, the universal Chilean bread. He stood inertly, staring at us, striking a self-consciously heroic pose.

The whole prospect now seemed more like a fairy-tale journey through a dark and mysterious wood than a well-informed sortie. Nevertheless all was settled. Tomorrow Eduardo and Pedro would bring seven horses for an early departure. Cárdenas would not accompany us, as he was leaving Mallén Grande with his family for his first vacation in three years.

After the conference, we looked forward to a full dinner and a good night's sleep. Over the meal Daniel shocked us with an announcement that he did not intend to accompany us on the ride. He admitted that the last horseback ride to Pico Sur and back, including his witness of my accident, had terrified him. He could not keep pace by walking, nor could he ford any of those raging rivers on foot. This was a blow, because none of us could speak Spanish fluently and the *baqueanos* knew virtually no English. Daniel tried to teach us some colloquialisms for the basic instructions and questions—tighten the reigns, tie the horses here, and so forth—but we were feeling a bit lost at the southern end of the world.

The next day, January 15, things further deteriorated. Eduardo showed up early but kept saying it would take a long time to shoe the horses. Gilligan arrived and offered the news that he would not be going with us and sheepishly pointed to his allegedly tender gut. As we had feared, this stylish cowboy was no cowboy at all; he was indeed more like a Gilligan than a John Bell Hatcher or the mysterious rider on the stallion above Pico Sur. Cárdenas, obviously upset by the erosion of the situation, offered to forgo his vacation and go with us, as his poor wife looked on anxiously. We would have none of this, however, and finally saw Cárdenas and his family speed off for a dependably late *Pilchero*. Eduardo claimed he could find another guide, but did not show up at the promised hour of 11:00 a.m. Meanwhile the rivers were rising.

There was nothing to do but wait. As I rested my legs in the sun some of the others paced anxiously back and forth in the schoolyard trying to grab some annoying clucking chickens. People came by in a flatbed truck and carried off a mountain of empty wine bottles, doubtless from a Cárdenas going-away party that we had apparently missed. At 2:00 p.m. Eduardo finally returned to our schoolhouse, with the strong stench of Pisco on his breath. He slurred through a disarming speech that described the unraveling expedition. Ojuay, our other intended guide, has demanded 50,000 pesos or $250 for his services. No way, I answered. It was too early for extortion, especially if we had no idea where we were going or how to get there. Eduardo didn't keep us from other bad news: he would not be able to get us five or six horses, perhaps only two or three.

There is a time in any expedition with indifferent success where one is moved to cut one's losses and be grateful for the mere experience of roaming the exotic. January 15 was certainly that day for the 1986 Andean Paleontological Expedition. We had lost our

Chilean companion and two guides, and the necessary horses had not materialized. Moreover, the route to the fossil beds and the reliability of our remaining guide, even in a a sober state, were not at all certain.

To distract myself from the inevitable, momentous decision, I followed Daniel to the peasant house of the Hermasillo family. As we entered we heard a small child coughing violently in a small room near the front door. I bristled with the thought that the advance we gave Eduardo would have been better spent on medical care in Coihaique than on booze. But the kitchen was a pleasant retreat, where several women—a mother and her daughters—were preparing a simple meal of *amasado* and *cordero*. As we sipped maté, we talked with Eduardo and the old patriarch of the house, Segundo, who spoke an elegant, rapid Spanish. Segundo described a secret crypt with *vieja huesos*—old bones—near "Pampa Castillo," a broad clearing bounded by a huge, castlelike butte. I could not follow much of this clipped speech, but I could tell from Daniel's widening eyes and rapt attention that the story was important. I had some glyptodont scutes in my hand, and when Eduardo's wife saw them she said, "Pampa Castillo." She had actually been there many years ago on an easier ride from the south along the Río Chacabuco. Apparently there were bones scattered all over the cliffs. Daniel and I looked at each other and smiled. We knew we could hit the broad side of a barn door—a butte 1,000 feet in elevation studded with fossil bones—with or without a guide.

Daniel ran and I limped back to our headquarters as the afternoon sun brightened the huge sunflowers in the garden around the schoolhouse. We brought news of Pampa Castillo back to our strategic conference. It was agreed all around that Andy and Mark, uninjured and reasonably skilled on horseback, should take the only available horses early the next morning with Eduardo, if of course he showed up. John and I would take a two- or three-day hike into the mountains in the general direction of the alleged fossil beds, hoping to sniff out a few accessible outcrops. Daniel had no taste for either strenuous endeavor and would stay close to the village, prospecting some ancient marine beds exposed near the lakeshore.

The next morning, much to our surprise and delight, Eduardo managed to show up a bit after 6:30 a.m., only a half hour later than his promised arrival. He had two horses for Andy and Mark and a skittish *pilchero*. We watched the riders head into the thick beech forest across the road. Soon John and I headed west on the dirt road toward Puerto Guadal and the higher mountains to the south. There were moments when the pain in my leg was excruciating, but I felt no more comfortable lying down than I did walking, and reasoned that it was best to continue.

To add to the discomfort, we were heavily loaded down with gear and extra clothes. The same mountains that battered us with cold and wind were now providing an unexpected heat wave. We became very thirsty, but our drinkable water supply was limited, and the many streams in the area coursed through fields populated with livestock. As a precaution, we drank the creek water through a "magic flute," a PVC straw containing a series of filters that guaranteed safe water under even more extreme conditions—the cover of the packet showed a man sipping through this contraption from what looked like sewage. For two days we traversed the road, crossed wild rivers over stalwart bridges, tramped on the spongy understory of thick beech forests, navigated hills crowned with annoying nettles, and camped high on the shoulder of a scrubby mountain with brilliant vistas of Cerro San Valentin and surrounding peaks. It was one of the most stunning natural excursions imaginable, but as paleontological prospectors, John and I were failing miserably. We only hoped that Mark and Andy would do better.

The outcome of the mounted expedition expedition was not known to us until after 9:00 p.m. on January 17. John, Daniel, and I were back at the schoolhouse, arranging the supplies and gear for our anticipated retreat, alert to any sound of the return of our comrades. Finally, we heard a low rumble in the forest, the sound of horses galloping with the aid of our very expensive horseshoes. A few minutes later they reached the schoolyard, a spectral visitation in the shadows. The horses indeed looked barely alive and the riders were not much better off. Andy and Mark asked us to help them dismount.

"We rode eleven hours!" Mark declared.

"So . . . ?" was the inevitable question.

Mark hesitated before replying, "Not much, really."

Eduardo promised to bring a *pilchero* tomorrow to convey our baggage to the boat, and rode off. As soon as Eduardo was some distance away Mark reached for a sack and dumped a big pile of bones on the table.

"Oh my God!" Daniel exclaimed as we shined a flashlight on glyptodont bones and the teeth and jaws of rodents, litopterns, toxodonts, and various and sundry other herbivorous beasts. These were finds that earlier bone hunters like Darwin, Hatcher, Simpson, and the Ameghinos had fervently sought.

"We struck it rich!" Mark said excitedly. "This is only a sample. We don't want Eduardo to know, though. He kept spying on us, probably for his friend Ojuay. There are some strange people back there."

We made a solemn pact to keep the news of the discovery limited to Carlos de Smet and Paul Raty, with the intention of returning next year to an untouched assemblage of dazzling fossils.

On the afternoon of January 18, we reached Chile Chico after a bad churn in a small police boat on a very choppy Lago Carrera. The wind sweeping down from Cerro San Valentin had transformed the lake into a stormy Antarctic Ocean. We staggered out of the boat at the *Pilchero* dock in Chile Chico and slept for thirteen hours in real beds at the Hotel Nacional. The next day we shared the triumphant news with Carlos. Paul Raty's father, a very refined, aristocratic gentleman, joined in our meeting, helping us immensely with the little-known geography around Pampa Castillo. We eagerly made plans for next year's expedition.

After a wild going-away party on January 20, we reported to the Chile Chico airport with severe cases of *cana malo*—hangover. Unfortunately, the flight would be hardly curative. The wind was still blowing very hard and our pilot was greatly agitated, imploring us to pack our gear quickly so he could get aloft before the wind became impossible. Carlos looked concerned, but he shrugged and smiled at me. Beatrice consoled us, "Don't worry. He is a good pilot, he is alive."

Within seconds of our takeoff, the plane pitched and rolled violently. Although we were buckled in, we knocked against the fuse-lage as our legs shot out from under us. Andy, who seemed most prone to airsickness, simply put a garbage bag over his head and prepared for the worse. I tried to look out the window, but soon resisted after we nearly scraped the belly of the plane on a rocky saddle between two peaks.

At last we made a steep descent into the more sheltered valley of the Balmeceda airstrip. At the end of the runway we noticed a chartreuse airplane with a skull and crossbones on the tail fin. Andy, now regaining his balance on his rubbery limbs, recommended, "Let's take the pirate airline next year, I hear you get free *maté* on the flight!"

By the time we reached Santiago late in the day we were feeling much better. At a carnivore's haven, a spacious *asador* called La Braziliana, we stabbed huge hunks of grilled sheep and cattle, and talked with great enthusiasm about the logistics of the future expedition to Pampa Castillo. This prominent feature is not indicated on any map, but Mr. Raty had a vague recollection of the place. He advised us next year to take cars down to the long road from Puerto Guadal at the far southwestern corner of Lago Carrera and make base camp at Valchaca, one of the Raty family ranches. From the ranch we could cross the Río Chacabuco on horseback, if the weather was dry and the river low, and then take an easier, six- or seven-hour trail a few thousand feet up to Pampa Castillo. Our starting point would then be located just a few miles away from the southern termination of this dirt road—all that was left here of the great Pan-American Highway that threaded its way from Alaska to southern Chile. Pampa Castillo was indeed at the ends of the earth; it overlooked huge, precipitous mountains to the south, the sentinels of the great Patagonian Ice Field that extended nearly to the southern tip of the continent. It was certainly the last outpost for fossils in this wild region north of the Ice Field. And, much to our frustration, it would be twelve long months before we would crawl over that massive, lonely butte for bones.

20

· · ·

Bone Hunters in Patagonia

I t was actually a few days less than a year, January 10, 1987, when
we pitched a camp in a stand of *Nothofagus* between the serene pas-
ture and the red rock castle of Pampa Castillo. This year we had a
slightly larger team, but it was still hardly a major expedition. Besides Mark, Andy, John, and myself, we brought along two other
Norte Americanos, Paul Sereno, a student with interests in dinosaurs
in the Columbia University–American Museum doctoral program,
and Roger Carpenter, a friend and colleague of mine from San
Diego State. Joining us at this high camp was Carlos de Smet himself and his young son Olivier, as well as Daniel Frazinetti and
René Burgos, our quiet, very professional *baqueano*.

In contrast to the chaos and uncertainty of our explorations
the year before, the 1987 trek to Pampa Castillo went remarkably
smoothly. We had rented cars out of Coihaique, where Carlos, the
ex-mayor of Chile Chico, and his wife Beatrice now lived. We

loaded these onto the hulking *Pilchero* for the passage across Lago Carrera, loaded off in Puerto Guadal, and drove to Valchaca, the ranch managed by Carlos and run by Paul Raty's brother, Jean-Luis. We then crossed a placid Río Chacabuco in the early morning on horseback and either hiked or rode the 3,000-foot climb up to the fossil butte. Travel from New York to our alpine fossil locality at the end of the earth—by plane, car, boat, horse, and foot—had taken a mere seven days. Moreover, our reunion had put Carlos and the clan in happy and high spirits. Only one tragic event marred the picture. The wind-tossed Air Don Carlos plane that in 1986 had bounced us down to Chile Chico and—with even more violence—flown us back to Balmeceda had crashed some weeks before on that very same route, killing our friend the pilot and three small children from Chile Chico.

As we pitched tents under the weather-beaten southern beech trees, I took in the panoramic mountain vistas around this lonely place. The massive cliff face of the butte rose over a thousand feet out of the forest. Its rocks looked more maroon and vermilion under the shade of an afternoon cloud than the red Pampa Castillo rocks back in the lab in New York. To the south, the fearsome glacier-clad spire of Cerro San Lorenzo was playing host to a team of world-class climbers. San Lorenzo shared the throne with Cerro San Valentin to the northwest and a jagged line of spires above the wild Río Jehmeneni to the northeast. Beyond the trees and the silver-green Pampa were more beech-covered hills that seemed to roll in endless waves toward the Pacific. Many of the summits seemed unassailable and the forests below them impenetrable, yet here in the middle of this wilderness was the best fossil locality for land vertebrates yet found in Chile.

But getting to the fossil-bearing layers of Pampa Castillo, even from our high camp, was no trivial exercise. The fossils were exposed not at the foot of the butte but nearly a thousand feet higher, up steep, crumbling slopes of hard siltstone and sandstone. That first afternoon, our climb and our brief bout of prospecting brought shockingly good results, including shiny blue-black teeth and chunks of mammal jaws. Fittingly, Carlos came back with the find of the day, a superb upper and lower jaw with a set of perfectly pre-

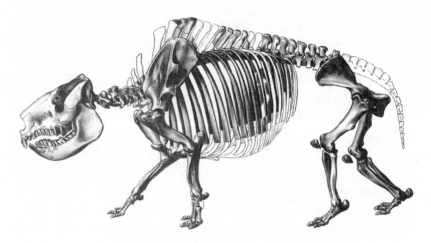

Another fossil mammal from the Miocene of South America: the lumbering,
rhinolike toxodont *Nesodon*

served teeth that belonged to a notoungulate. That evening we
talked excitedly about anticipated discoveries, relishing portions of
the fresh, early-season barbecued sheep.

The next morning, after a breakfast of hot coffee and greasy *tortas*, we faced the ascent of the butte in the howling wind. My legs
were not altogether fit after the Andean horse accident of last season, and I grimaced in pain whenever I had to flex a knee to scale
a boulder or ledge. Near the top of the butte, the sandstones and
siltstones began to show bone in isolated spots. As the fossils got
better we got spoiled. Andy found an entire skull with jaws of a
big rhinolike toxodont, well encased in a greenish gray sandstone
ledge.

Many days of great paleontology followed. We scrambled all over
that big butte, extracting skulls and skeletons of numerous strange,
herbivorous, camel-like, antelopelike, and rhinolike mammals, as
well as armored glyptodonts and rodents. The census of our fossil community revealed that these animals were indeed probably
Miocene in age—about 20 million years—and very similar to those
collected by Simpson, Hatcher, the Ameghinos, and, before them,
by Darwin himself. There were no big surprises, but it was quite
satisfying to recover this Miocene fauna so far west and so much
higher up than its usual localities on the Patagonian plains.

We called our giant butte with its fossils Cerro Smet, in honor of our happy Carlos, who with his son Olivier had to leave after only a few days to get back to his business in Coihaique. We continued our collecting, intending to try to hold out at the site as long as the weather would allow, through January if possible. There is true contentment in working such a productive fossil site. Each day of persistent hiking and quarrying brought its own special treat—a fossil of an animal we had not previously encountered. Within a week we found a real gem—a tiny jaw and teeth of a rare, opossumlike marsupial. But there was a prize in every Cracker Jack box, and spirits were high. We would return from ten hours on the outcrop, shivering cold and wet but happy with success, ultimately restored with a *bombea* of hot maté and a fire stoked with rotten beech logs.

Despite our adaptation to the conditions, the changeable weather started wearing us down. Sunday, January 18, was so hot and windless that it made us think of Baja California. January 19 began with a snowstorm. The food rations, of questionable quality, were also rather debilitating. The fried bread that René so faithfully prepared for us every morning began to take on a rancid, waxy taste, as the mutton grease solidified on the roof of the mouth or coagulated in a lump in the stomach. We developed the habit of impaling these *tortas* on a stick, like marshmallows, and burning them to a crisp over a fire, rendering impressive amounts of fat. Worse yet was the mutton itself. The cold weather of Pampa Castillo kept the carcasses reasonably preserved, but after the brief warm spell the meat took on a greener cast. We cooked some tough chunks with the marshmallow stick method and watched the maggots try to escape from the inferno.

On January 22, a day of fair weather, Daniel, Roger, and John packed up to return. The skeleton crew of five who remained planned to work for several more days, hoping the weather would hold for a safe departure. It was not to be. Late on the 24th there was a feeling in the air of an approaching cold front. René looked anxiously toward the west and north. We retreated to our tents, and I lay there in my own golden nylon dome, listening to the BBC on my shortwave radio. A lovely Brahms violin sonata drifted above me, intermittently obscured by radio static or the wind that shook the trees above.

It rained torrentially from eleven that night until ten the next morning. I was roused continually as the wind broke off big branches of beech or shook fountains of water from the trees onto my tent roof. Even after the rain diminished the ground was impossibly muddy. We spent the 25th mired in a camp transformed into a disaster refuge, with drooping shrouds of nylon, piles of felled rotten logs, and a fire that could release only a wisp of smoke. On the 26th we made a desperate attempt at a last day of honest fieldwork. Alas, the weather was not cooperative; the storm was returning. Wrapped in piles of down, Gore-Tex, and nylon, we tried to work Bangle's Quarry, a small site on the southeast edge of Cerro Smet, where we had found several rodent skulls. After scraping away frozen mud as the icy rain beat down on us, we at last left the site for camp. Paul and I struggled for an hour to start an emaciated fire. That night we scrounged around for a few rations and finished off our coffee, very ripe cheese, and a few bits of chocolate. Then a powerful storm again engulfed us.

Back to my tent. I became familiar over the next few hours with every word of a *New York Times Book Review* section dated December 28, 1986, wherein I learned some quotable quotations. Flaubert, writing to Turgenev, advised, "Describe a tree so no other tree can be mistaken for it." It seemed good advice for plotting out a strongly supported evolutionary tree of relationships among ancient South American mammals, dinosaurs, human lineages, life itself. A book reviewer asserted, "If we write about our scientific observations, our observations inevitably color our writing." I tossed this phrase around a bit, and it began to annoy me, as if the words were put together to seem more important than the idea actually was. This mental self-torture persisted as I became bored and claustrophobic; the whole orb above me was the dull yellow of a tent dome pounded with sheets of rain.

The morning was no better. I escaped from the rain and the mist at about six and trampled through the mud and the moss. The situation was very bad. No river in southern Chile could be safely crossed under these conditions. We had often heard stories of *baquea-nos* being marooned in the mountains in such a storm or drowning along with their horses in a desperate attempt to ford a flooded river. I sloshed through a swamp, in some places knee deep in water that formed a moat around René's ragged tent. To my morning greeting he poked his head out of a mass of pelage, a thick pile of sheepskins and wet dogs.

"René, *vamanos?*"

"*No, río peligroso,*" he answered, commenting on the swollen torrents that blocked our path.

For nearly an hour we huddled under a tarp around a fire pit, desperately trying to induce a flame from soggy wood, ash, and garbage. René paced back and forth, anxiously looking toward the direction of the wind from the west. The sky on that horizon was a dark green.

Suddenly he said, "*Listo, ahora!*"

So we were going after all? We looked at each other in confusion but followed René's command. We spent a miserable time digging out a pit for the garbage that kept collapsing in the rain-soaked peat. We then wrapped the tents, wet and slimy like dead eels and smelling of old clothes and stale sweat. There were not enough *pilcheros* to carry the gear, so we left a cache of equipment (bottles, picks, cooking gear) for René to pick up later. It took René two hours to load up the temperamental packhorses in the driving rain.

So now I was again forced to take a dangerous ride on a horse, a means of transportation for which, even under normal conditions, I had lost some fondness. My horse bolted the moment I mounted; I checked it with a sharp tug of the reins. The rain was a thick curtain across the open clearing and the horses made a fast clip for the more sheltered grove of trees at the far end of the glade. These trees also marked the trail for the descent. We were soon in a ghostly forest dripping with rain, where low branches fringed with Spanish moss themselves looked like shaggy tree sloths. We had to duck low under many branches to clear them. Andy warned us about a particularly tricky section of the trail where, during the ascent, he and his horse had been separated, Robin Hood style, by a massive, low-hanging branch. As in our retreat of 1986, only the skinny dogs—the "guitars"—seemed to be having fun. They ran out from the pack, disappearing in the mist after an agile fox, a descendant of those mammals that invaded South America from the north three million years ago.

Our emergence from the dank forest hardly brought relief. The trail was now a narrow ribbon perched over steep muddy slopes that plunged down to swollen cataracts. The terrified horses wanted to exit this place as fast as possible, and opportunistically planted their hooves in the mud and slid down in that sickeningly vertiginous manner I remembered from the retreat from Pico Sur. This pattern of walking the tightrope of narrow wet ledges and sliding down slopes finally ended in a little over four hours. It seemed we were out of the worst of it; the terrain was rolling and grassy under an occasional sun. I began to convince myself that even the Río Chacabuco could be negotiated.

This optimism was of course completely unjustified. When we reached the Chacabuco, René gasped in amazement and laughed. The peaceful river we had forded only three weeks ago had become a raging torrent a quarter mile wide. Rapids with standing waves several feet high seemed to carry all the flotsam and jetsam of the southern Andes—huge beech logs, the siding of a former riverside cabin, the bloated carcass of a cow, and countless other items. I asked René how long it would take the river to fall to a navigable level. *"Tres días,"* he murmured and shook his head. He knew that forecast held only if the weather vastly improved. He also knew this was a problem for us. We would surely miss tomorrow's ferry at Puerto Guadal and would thus be stranded in that forlorn port an extra week. The delay would cascade into a disruption of important plans for meetings in Santiago and our scheduled flight home.

"Well, we're stuck here, that's the way it goes," Mark said several times.

"Yeah." I shrugged my shoulders. So what's the big deal? We all made it safely down the mountain. There could be worse places to be stranded. We walked over to a small cabin set safely back from the river, where an old rancher fried us some *tortas*. Nice, I thought, maybe I'll just settle down by the banks of the Chacabuco, which in its more normal condition was a stately stream of brilliant blue-green, glacier-fed water, swarming with big salmon.

But René was not at all happy with the situation. He was professionally devoted to us, and felt he had taken a risk, despite the impending weather, in allowing us to work a few more days. He had the sad look of unjustified shame and failure. We tried to console him with expressions of gratitude and contentment, but he would have none of it. Suddenly he jumped up and walked out of the cabin. He then tied his dogs with a hefty rope to a post and mounted his horse.

"He's goin' for it!" Mark exclaimed. I couldn't believe it.

René reasoned that a river that could drown a dog or a human on foot with dispatch might still have a line of weakness for a mounted crossing. We watched him pull hard on his horse and practically disappear in the maelstrom, then reemerge back on our bank of the river. He made these vain sorties for over an hour, moving up and down the river. Clearly the water was too deep. A horse unable to reach bottom and clear its head is no better off than any other carcass—man, dog, or cow—floating at breakneck speed down to the Pacific.

Then René rode quite far, about a mile down the riverbank. As we approached this spot we could see he had made it about halfway across. The horse carefully weaved a path from one submerged bank to another. It would emerge on a shoal, dripping with mud and icy water, then suddenly plunge into the next channel up to its neck. But the river was quite wide here and the channels were a bit shallower. It took René fifteen minutes to make the slalom to the other bank. He immediately turned the skittish horse around and returned across the river, confidently retracing the successful path.

"Vamanos!" he shouted, with more urgency than triumph. The river was still rising.

What followed was a horrible fifteen minutes of semiaquatic horsemanship. My heart pumped wildly as we plunged into the first torrential channel. Even above the roar of the river, I could hear the nervous grunts of the horses and the sound of rocks cutting loose from their moorings on the river bottom. René crossed each deep channel and turned right around on a bank, prepared for a rescue attempt in case a horse lost its balance. But we all made the crossing. On reflection, one might question René's decision to put his clients at such high risk. But this was, after all, a frontier, and he was our hero. He successfully brought us back to Valchaca in one piece and on schedule.

The southern end of the riverbank was a paradise. A warm sun started to dry our sweaters and anoraks, and we brought out some fresh cheese and bread from a ranch house. With much gratitude we said our farewells to René with promises to return in 1988, and boarded our rental cars for the scenic road trip back to Puerto Guadal. Unfortunately, one of the vehicles, an old Datsun pickup truck, was rather lame and broke down several times. We said unkind things about Fidel Pinella, the obsequious fellow in Coihaique who had rented us this lemon. After a series of emergency repairs we ultimately got the Datsun to Puerto Guadal by evening, but we were dumbstruck to find there was no ferry waiting for us. We had risked our lives to cross a river on nature's beasts of burden, only to miss the ferry because of some poorly maintained machinery. Apparently we had been misinformed, and to reach the ferry the next morning in Mallén Grande would require another twenty miles' drive in the dilapidated Datsun.

The rest of our journey home was without incident. We carefully packed the fossils for transport in our luggage in the airplane and talked about plans for next year's expedition. News of our success had preceded us, and back at the museum in New York I was assaulted with questions about our discoveries from colleagues, visitors, and former professors. As noted, the animals—the glyptodonts, bulky toxodonts, hoofed, long-legged notoungulates, marsupials, and rodents—were familiar components of Miocene-aged faunas from Argentina. But this was the first time such a classic assemblage had been found at a 10,000-foot elevation deep within the Chilean Andes. Marine rocks just below the bone-bearing sequence indicated that this was an ancient shoreline, one that had seen a remarkable rise in elevation in the last 15 million years. Indeed, our work at Pampa Castillo supported the theory that the

Andes were among the fastest-growing mountains in the world. *The New York Times* went a bit far with this connection in an article headlined "Whale Fossils High in the Andes Show How Mountains Rose from Sea." A letter to the editor that otherwise gave much credit to the find wryly asked, "Where else would they have come from?" Nonetheless, our discoveries were important on many fronts. Perhaps most significantly, they showed the exciting potential for unearthing fossils of ancient mammals, dinosaurs, and other land vertebrates in little-known and unlikely territory. A mountain chain several thousand miles long awaited further exploration.

21

Above the Clouds and the Condors

I n 1988 we returned to Chile with the intention not only to keep working Cerro Smet and Pampa Castillo but also to do some far-ranging reconnaissance in the Andes farther north. We could see on geological maps that several deep rivers cut across the range and flowed into the lush agricultural central valley of Chile south of Santiago. Indicated on those maps were some enticing swaths of blue that represented Cretaceous rocks, possibly with dinosaurs, and red and brown that indicated Early Tertiary rocks, possibly with fossil mammals. Daniel told us there was even a place with some magnificent dinosaur trackways in a cliff near a resort with the curious name of Termas del Flaco ("hot baths of the skinny man").

But before we traveled north, we took another pass at Cerro Smet. By January 12 we unloaded our Fidel Pinella rent-a-wrecks from the *Pilchero* at Mallén Grande, where we again spotted sinister Pedro Soto, alias Gilligan, this time with smaller spurs. By late that evening, we reached Valchaca and reunited with our reliable *baqueano*, René Burgos. Again we drove a reluctant pack train up to our desolate butte and settled in for a siege of its fossiliferous rock walls. And again the awful weather conditions of the Patagonian Andes reminded us of the discomforts of field life. We worked the site from January 14 to January 25, only twelve days, nine of which were filled with gales, snow, and rainstorms. Bad weather that set in on January 14 produced enough snow on the 17th to coat Cerro Smet and the surrounding buttes thoroughly, obscuring any indication of fossils or other surface features. An entry in my field journal fairly sums up much of the experience. "1-19-88. Mark and I worked Bangles Quarry. Miserable day. Snow, rain, sleet. Felt like shit. Found 3 rodent skulls, various jaws and teeth. . . . Worked until 7:00 p.m. Feet numb."

For all this inconvenience our haul was very good, if not as spectacular as that of the year before. It was clear how much difference the weather made. Despite our discomfort the previous year we had many more days of sun and workable conditions, and the accumulated man-hours on the outcrop far exceeded ours for 1988.

Back in Coihaique we explained to Carlos that we were going to pursue other options in the mountains to the north, but this did not necessarily mean we were deserting his scenic region. We tried to reassure him that we still believed that miles of terrain around Valchaca were well worth prospecting. Carlos was not consoled, however. He knew we were not content with a Miocene assortment of fossils that was hard to distinguish from the classic assemblages worked for decades in Argentina. He kept talking about places nearby where there might be dinosaurs or more ancient, pre-Miocene mammals, and he promised to arrange for helicopter travel for better and more efficient access. We feigned some interest, but were itching to explore places where such finds seemed more likely. After a warm farewell to Carlos and Beatrice we stuffed ourselves into one of Fidel Pinella's lousy rental vans and shouted back, *"Hasta luego!"* Unfortunately, after more than fifteen years, I have yet to see Carlos and his lovely family again.

Part of the reason for this long separation from the de Smet family was our surprising and distracting success in the north. Early in that 1988 season, this outcome was by no means auspiciously indicated. Indeed, we had only a half-baked plan for the paleontological reconnaissance of the Andes. We would fly the several hundred miles north from Coihaique to Puerto Montt, where we would rent a sedan and jeep and take a motor trip for 350 more miles along the coast to the large industrial port of Concepción. From that point we would head inland and drive up one river valley after another, looking for likely outcrops, whose close inspection would doubtless require a rugged horse ride or scramble on foot. Our maps were sketchy and large-scale, without sufficient details of landforms, topography, drainages, and roads. To make matters worse, we lacked the help of a local expert on geography. Daniel Frazinetti had worked extensively in Chile north of Santiago, where marine rocks with fossils were plentiful, but was barely more familiar with our destination area than we. Any success to this capricious venture seemed extremely remote.

Our stay in Puerto Montt, the base camp for this tentative survey expedition, was marked by two events that revealed Chile at its best and its worst. Puerto Montt is a bustling fishing center with an intimate attachment to the great Pacific. Perhaps nowhere else in the world, even in Japan, is there a greater devotion to seafood. On the docks, tiny restaurants, each affording seats for no more than a dozen diners, were jammed together like stalls at a street fair. The cuisine in these establishments was derived from the fish, crabs, clams, and other food items dumped directly from fishing nets that day onto giant wooden tables in the center of the docks. Daniel took us to his favorite fish shack, number 82, where two smiling middle-aged ladies gave us *teecita*, red country wine poured into tiny teacups. The ladies then proceeded to engorge us with the kingdom of the sea; rather than treating us with selected delicacies, we were presented with whole ecosystems. A large soup bowl, *pailla marina*, looked like a tidal pool with small fish, clams, squid, and even tunicates. The feast included *congria frito*, a delicious, lightly battered fried conger eel; *chulgas zapatos*, giant mussels; *choros*, small mussels; *almejas*, clams; *centollos*, king crab; *locos*, abalone; *picorocos*, barnacles; *jaivas*, more crabs; and *curanto*, a pot with a ton o' fish. I had never tasted such fresh, unadulterated marine food in my life.

In contrast to the maggot-ridden sheep and greasy *tortas* of past days, this was dining pleasure of shocking intensity. At the end of the banquet, we rubbed our distended stomachs and surveyed in amazement the huge pile of *ressaca*, beach trash, on the table.

The other spectacle in Puerto Montt was far less edifying. When we took an evening stroll into the center of town, we passed a large gathering of demonstrators wearing red armbands and waving red flags embossed with gold hammer and sickles. With placards and speeches they loudly protested the recent disappearance of five young students from the nearby university, suspected causalities of General Pinochet's crackdown on political dissenters. Then, when we reached the main plaza, there was not even a hint of anything politically unusual. From a small bar with pinball machines and a big picture window, we watched shoppers and tourists cheerfully making their rounds outside. Soon, however, we heard shouts and whistles as the demonstrators migrated toward the plaza. The proprietor of the bar sprang into action, locking us inside with an elaborate cage and giving us little choice in the matter.

The reason for this extreme response was soon apparent. Squad cars and jeeps loaded with police, *carbonaros*, and antiriot equipment appeared in the street, rolling slowly but deliberately toward the demonstrators in the distance. These war machines were followed by a regiment of teenage antiriot troops, boys decked out in army green, wearing curious black Darth Vader helmets, and holding bulletproof shields. Most of them carried clubs and some of them automatic rifles. All of them looked nervous and scared. We sat in the bar for over an hour, wondering what tragic events were transpiring outside. But no gunshots, screams, or violent explosions were heard. The demonstrators dispersed and the antiriot troops retired to the north side of town. With the return of tranquillity, the proprietor unlocked the bar and allowed us back onto the streets to join the other *turistas*.

With little reluctance we left Puerto Montt on January 30 in a rented Chevy Monza and jeep, and drove along a coast incised with inlets searching for likely fossil outcrops in exposed canyons or road cuts. I had an unwelcome familiarity with the scene, for it reminded me of those scrambles for fossils among the road cuts and housing tracts of San Diego. We reached the pleasant town of Temuco, which displayed the strong cultural influence of the Mapuche Indians. Temoco and its surrounding countryside was one of the few regions in Chile where indigenous tribes managed to hang on, though in vastly reduced numbers, in the wake of European conquest. The next morning Daniel announced that he felt so useless that he had decided to return to Santiago and left us to wander about on our own. Reluctantly, we took him to the crowded Temuco bus station. It was clear that Daniel had much affection for us but little regret in cutting short his 1988 field season.

Daniel's enthusiasm for departure turned out to be highly explainable. Our route eventually took us along the seacoast between Lota and Colonel, two industrial coal-mining towns whose activities had polluted the atmosphere, the land, and the sea. Smokestacks that seemed as tall and phallic as the volcanic spires around Chile Chico spewed sulfurous clouds that retinted the sun a dull orange. Huge tankers floated offshore in an oily ocean fed by toxic outflows from the factories. This seemed a poor place for the survival of any life-forms. Piglets covered in soot, each sporting a curious forked stick around the neck to prevent them from passing through a fence, ran in panic from squadrons of nearly naked street urchins. We traversed this miserable stretch of shoreline, referred to euphemistically in our guidebook as "the Carbon Coast," and headed toward our hotel in Concepción.

When, some days earlier, we asked Carlos de Smet whether Concepción was a nice town, he paused pensively for a moment, then smiled in his customary way and replied, "No!" The slums along the Río Bío-bío were as devastating as any in Mexico, but the center of town and our hotel were pleasant enough. Besides, we knew that the Bío-bío, one of the great cascading rivers on earth and a mecca for extreme kayakers, flowed directly out of the Andes, where we would soon be again. One more day of driving inland and then 250 miles north through a central valley filled with orchards and cornfields brought us to the foothills of the mountains.

In the busy agricultural center of San Fernando we tried to follow up on some sketchy rumors about fossils. Allegedly some dinosaur bones had been found near a mine west of town, but we couldn't make contact at the phone number provided. We turned our attention east to the more likely-looking terrain of the high Andes. Termas del Flaco and its dinosaur tracks lay some fifty miles east on a winding road. We were hopeful—if dinosaur tracks, why not bones nearby?

Access to the Termas road was not automatic. A guard station bristling with guns and munitions marked the restricted entrance. The officials at the station informed us that the narrow road allowed only one-way traffic and the road was closed to our direction until 4:00 p.m. As he provided these instructions in a cordial but clipped delivery, the officer was accompanied by a one-man chorus. A drunk locked in a cell at the back of the station gave out a thunderous aria, sung with Pavarotti-like passion but not skill. A long line of pickups and battered sedans started to queue up for the eastward passage to the mountains. I took a leisurely stroll near the guard station, but my meditative walk was rudely terminated by some excited soldiers. Apparently I was walking over a minefield. Why all the fuss and security? I wondered. The officer informed us that in the next canyon south the guard station had been bombed by anti-Pinochet subversives, killing several of their troops. The pleasant, orchard-filled valleys of Chile could not obscure the oppression and retaliation that plagued this beautiful country.

The fifty-mile drive to Termas was a flight to an alpine heaven. The road made switchbacks up a huge escarpment for several thousand feet and then plunged down into the valley to serve as the left bank of the roaring Río Tinguiririca. As we gained elevation, dense scrub and woodland gave way to stands of pine and finally alpine tundra. Termas itself, a cluster of small hotels around extensive hot springs, was nestled in a huge crypt bordered on all sides by peaks ranging from 11,000 to 18,000 feet. The bottom of the valley, where the town and the hot springs were located, was warm even at seven o'clock, the hour of our arrival, but high above us a fresh coat of snow blanketed the summits and fed the glaciers and snowfields.

Then we caught sight of something very unusual. The great walls of the canyon below the snowy peaks were formed from red, brown, and tawny sedimentary rocks, some swirled like chocolate,

some tilted so vertically that their flat planar surfaces were fully exposed as huge slabs, some exquisitely layered and tilted in directions
that cut across the grain of other rocks nearby. The whole hodgepodge of exposures extended up to a stupendous height; it was like
a Grand Canyon tilted on its side. At the top of the 5,000-foot escarpment south of the river was a mass of folded yellow sedimentary rocks forming a golden throne. Just west of this peak was a
summit of purple-gray tilted rocks fringed with big boulders and
snowfields. The top of the escarpment in that area was marked Cretaceous on the maps. "That's where the bones are!" we all practically shouted in unison. But a prodigious climb would be required
to get to this or any other likely spot high above the valley.

A tough ascent the next day called for a good night's sleep, and
we stopped at the first place in town, the Hostelería El Rancho.
The architecture of this resort haven was a strange mutation, com-
bining Mediterranean and Swiss chalet design, complete with
flower boxes. The proprietor of the establishment was less than welcoming; he gave us a look of suspicion and glared at us as the
young waitresses stared and smiled. We were not his usual clients—
older, respectable, clean-cut Chileans. Termas del Flaco was not an
international Andean resort like the ski center of Portillo; there was
no real desire here to attract foreigners. Besides, we were dirty and
unkempt; we were "the creatures." O citizens of Termas del Flaco,
lock up your doors and windows, protect your daughters! The
creatures have arrived!

The maid shared no such hostility for us. She stepped brightly
into our large bunkroom and asked me to help her raise the mirror
in the bathroom, proclaiming with approval, "*Norte Americanos* are
so much taller than Chilean men!" After cleaning up as much as
possible we went to the dining room for a bland meal of chicken
and string beans. Here it became obvious that we were expected to
be more than boarders; the high price of a night's lodging signalled
that some obligatory entertainment and social mixing were scheduled, as if we were a tour group. The room was full of mainly older
couples, laughing and clapping while a strolling guitarist assaulted
us with Chilean folksongs and lounge-act favorites like "Danke
Schön." At one terrifying moment the guitarist tried to play

matchmaker, urging us to share a Chilean traditional dance with ladies seated at a nearby table. We were rescued by some enthusiastic Chilean men, who partnered the women; the dance consisted mainly of waving handkerchiefs. The whole ritual ended precipitously when the intrusive guitarist accidentally scratched his precious instrument against a chair. Cradling his guitar like a wailing baby, the entertainer ran outside to inspect the damage in the better light on the veranda. His exit was the cue for "the creatures" to retreat to their rooms while the other guests congregated on the porch for more socializing. Nothing is worse than obligatory fun.

The morning weather on February 3 was superb, clear and windless with a dark blue, almost violet sky, a good day for a long climb. At breakfast we learned from one of the more affable employees that they had been instructed not to talk to us. Beware the creatures. Nonetheless, in cheerful defiance the staff treated us very well and packed lunches for our jaunt. A twenty-four-year-old geology student from Antofagasta, who called himself Max, came by, reminding us that in an encounter in town the evening before we had invited him on our scouting trip. Over breakfast we decided on a two-pronged assault on the mountain range. Mark and Andy agreed to climb the canyon walls to the south to check out that enticing set of purple-gray beds at the top. John, Max, and I would stay north of the river, inspect the slabs with the dinosaur trackways, and look for any nearby tracks or fossils. Then, like Mark and Andy, we too would take to the hills, climbing to the top of the northern escarpment. Unfortunately, this prospect was less alluring than the southern route. From our position deep within the canyon we could not see what kinds of rocks were exposed a few thousand feet directly above us.

The dinosaur trackways turned out to be well worth a visit. They were not only spectacular but also highly unusual in their exposure. Typically, trackways were formed when the animals walked across the muddy bottom of a lake or a stream, leaving marks that became impressions in horizontal beds of ancient limestones. But the tumult and twisting of rocks pushed aside with the rise of the Andes had tilted the limestone slabs at a very steep 70-degree angle. When we rounded the corner of the hill some thousand feet above the town we gasped in amazement. It looked as if these giants had zigzagged straight up the side of a mountain!

The diversity of tracks indicated some diversity of dinosaur species walking across the once-horizontal bottom of the lake. But the dominant and largest tracks were made by long-necked, thunderous sauropods. Each footprint of these monsters was a sizable pit a couple of feet in diameter. We scaled the steep wall aided by curious hand- and footholds, the tracks themselves. Nearly one hundred feet up the cliff, I rested, sitting back into a track and dangling my legs over the ledge formed by the impression of the back of the dinosaur's foot. I peeled an orange and enjoyed the mountains, the deep maroon canyon, the ribbon of a river far below, giant condors circling far above, and scores of dinosaur trackways moving upslope toward me. How marvelous!

We ascended the top rim of the slab with the dinosaur trackways and continued working our way up the huge precipice toward the summit ridge. After a few hours John decided he had had enough. "I'll do the traverse prospecting," he offered. "See you later." I moved upward more briskly, leaving Max some distance below, anxious to find something of interest before the waning sun forced a descent. In the talus of a steep cliff I found some very hard dark brown limestone sprinkled with specks of fish vertebra. Very good, I thought, and kept moving upward. Only a hundred feet below the snowfields that fringed the summit crags I found a productive layer of rock, full of fish bone and gar scales. The chink of my rock hammer echoed through the canyon as I worked energetically to remove a nubbin of fossil-rich rock welded to the mountain face. After some strenuous chopping in the thin air, I at last had a sample worth bringing back to the hotel. This puny extract at least showed the potentials for vertebrate paleontology at 13,000 feet in the Andes.

Farther downslope John had also found some fossil bone. We were pleased with ourselves, but our friends again scored much better. Our hunch about the auspicious-looking beds that capped the southern escarpment was dramatically substantiated. Forty-five minutes after we returned to El Rancho, Mark and Andy showed up with an armful of rocks full of extraordinary mammal skulls, jaws, and limb bones. These were rather weird forms, much too primitive to be the Miocene animals we had extracted down south from Cerro Smet, but too young to be Mesozoic and therefore contemporaneous with dinosaurs. Oddly enough, only Mesozoic rock had been mapped on the crest of the range, so we had exposed a scientific puzzle well worth solving. Mark noted that the extensive, tilted red beds below these purple beds were probably Mesozoic though, and they looked good for the possibility of dinosaurs and other reptiles. We were stunned by our streak of good luck in the Andes. The first probe up a deep river valley and the first prospecting on the first day in the region had revealed a dramatic, previously unknown assemblage of bones. The mission for 1988 was accomplished.

We were of course eager to charge back to the southern rim over Termas del Flaco the next day, but this was impossible. National and regional official clearance was required to set up a long field program, and the necessary discussion and paperwork would delay our effort until the 1989 season. But we were confident that such an unexpected discovery would attract the necessary funding to return to the recuperative hot springs of the skinny man. Before returning to Santiago, however, we did again try to track down those tales of dinosaur bones in the valley west of San Fernando. This required a twenty-five-mile drive west to Santa Cruz, a pleasant village surrounded by vineyards. Here we met a woman who belonged to the family owning the land and the mine where the dinosaurs were supposedly found.

The results of this reconnaissance were disappointing. The rocks of the mine were Cretaceous marine layers, and the bones—long since destroyed or reburied—were probably fragments of the skeleton of an ancient marine reptile, not a dinosaur. Indeed the excursion would be hardly worth mentioning if not for a curious and tense encounter. The women left us alone to explore the mine and the deep quarry nearby. As we probed deeper into the mine we could hear voices. Peering around a corner of rock we could see a group of soldiers playing cards. The mine echoed with the sound of poker chips clinking on the cave floor. We could see the men circulating a very large keg of wine. We went about our business of prospecting, but this turned out to be a rather impetuous mistake. At the first clang of a rock hammer, the men sprang to attention and a very aristocratic officer mounted his horse and commanded us to return to the entrance of the cave. By the time I reached the cave entrance, the officer was miraculously already there, interrogating John with marked intensity. The *carbonaros* now looked much more formidable, standing stiffly at attention and clutching their short-barreled automatic rifles. As I stood waiting for my turn in the interrogation line, I caught sight of our hostess's car, screaming back toward the mine ahead of a great tornado of dust. When she reached the mine the woman jumped out and started talking in a very agitated manner with the officer. She carefully explained the reason for our presence and apologized for spooking the soldiers.

Our encounter with these sinister soldiers was a riveting reminder of the uneasy state of the country. In 1988 as in previous years, we could sense the political vise grip of the Pinochet regime and its tenuous hold on various factions. Retired officers who ran small hostels proudly displayed photographs of Chilean battleships. A very popular image was a photomural of the bombing of the Presidential Palace (while Salvador Allende, the democratically elected president, was still inside) that had signaled the killing of 3,000 people and the ascent of Pinochet to his dictatorial position in September 1973. Meanwhile thousands of students were demonstrating against him in the streets. Santiago was vibrantly busy by day but mysterious and foreboding at night, when a 10:00 p.m. curfew left the streets to the *carbonaros* and their guns. The apparent involvement of some U.S. organizations—perhaps the CIA—in all this was not pretty. When Senator Edward M. Kennedy paid a visit to Santiago, the buildings and street posts were plastered with posters of Mary Jo Kopechne, the woman who had drowned in Kennedy's car after an ill-fated party in July 1969. The expertly produced signs said in English, "Remember Mary Jo." This was clearly not the work of local Chileans devoted to the cause of dictatorship. One could envision some U.S. intelligence team conveniently providing such signs to Pinochet and other dictators— fierce opponents of the communist peril—around the globe. We had witnessed the riots in Puerto Montt, the bombed guard station, the land mines, and the soldiers in the Santa Cruz mine, whose furtive and suspicious behavior could only suggest some dark and perhaps even murderous purpose. As we drove away from the wilderness of Patagonia toward the gravitational pull and the intrigue of the capital city, Chile became very creepy.

Back in New York, another kind of Chilean political intrigue haunted us. Some time after our return I received a note accusing me of illegally sacking out and exporting a large mass of fish fossils from northern Chile. This was an outrageous case of mistaken identity, but one I was forced to confront. Our expeditions had not been within a thousand miles of the controversial locality and I had the documentation to prove it. Some of our colleagues in Chile theorized that a group of commercial fossil collectors had posed as a legitimate scientific research group at the locality and used my name in the process. Accusations were distributed to government research foundations and even to the president of the American Museum, a situation that did not greatly please me. After an exchange of letters the matter was clarified in my favor, but not without several months of stress and strain.

Unfortunately, when we arrived in Chile in January 1989 we soon learned that not all the intrigue had evaporated. We were told that there might be a news story and editorial in Chile's national papers criticizing our activities. A competing research party tried to keep us out of the Termas del Flaco area and suggested we confine ourselves to Patagonia. But we hung tough, intent on returning to the rich site below the summit ridge to which only we had a rightful claim. Paleontology continues to be a highly competitive and sometimes antagonistic pursuit, in the questionable tradition of the battles between Cope and Marsh in the dinosaur territory of the western United States more than a century earlier. South America seemed if anything even more likely to foster such struggles. As we packed our gear in the hotel I ruefully recalled all the accusations, letters, bureaucratic hassles, political intrigue, and backstabbing that had brought us to this moment. I was eager to get away from people and escape to the mountains.

On January 5 we arrived in Termas del Flaco, this time choosing a simpler establishment, the Hotel Gloria, for our lodging. The next

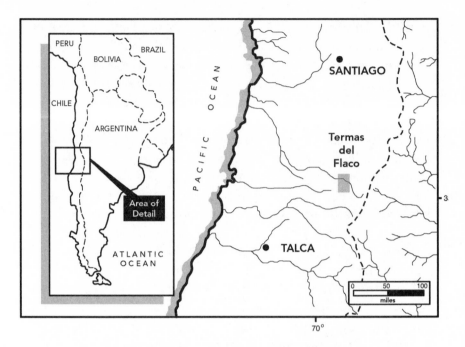

PERU
BOLIVIA
BRAZIL
CHILE
ARGENTINA
Area of
Detail
ATLANTIC
OCEAN

PACIFIC OCEAN

SANTIAGO

Termas
del
Flaco

3.

TALCA

0   50   100
miles

70°

Important fossil localities in the Andes of central Chile

clear morning we were on the lower set of outcrops, the nearly vertical red and coffee-colored sandstones of the Cretaceous Colimapu Formation. Some time between our departure in 1988 and our return, a group of geologists, inspired by the report of our success in this area, had prospected these beds and found an impressive Cretaceous lizard. Our foray on the steep rocks was only mildly successful, with John recovering a small string of vertebrae sealed into the hard rock of a ridge. We resolved to turn our sights the next day to the summit of the canyon, where the year before Andy and Mark had made their great discovery.

So began a gloriously productive project 4,500 feet above Termas del Flaco. A steep, arduous climb on January 6 brought us to the tilted layer of rather odd purple-gray rocks, which looked like mudstones mixed in with lava from a volcanic eruption. The layers weathered out into spherical concretions the size of baseballs, cannonballs, and bowling balls, concretions with wonderful properties. A large number of them had simply formed around unusually well-preserved skulls, jaws, and skeletons of various mammals—weird-looking slothlike forms, diverse notoungulates, even marsupials. In a matter of only seven days—from January 6 to the 12th—our team recovered 149 specimens, many from the cannonball concretions.

The setting for our productive labors could not have been more majestic. This was the rugged spine of the Andes where the mountains become ever more lofty in the northerly direction, reaching their apogee at 22, 841-foot Aconcagua. Across the valley and over the first bastion of snowcapped peaks was a remote snowfield that looked like an isolated white island in the clouds. We were told that this was the basin where several decades ago the plane with the Uruguayan national soccer team had crashed. The dark stories of the survivors, who were driven to feeding on the remains of those who had died in the crash, were made legendary in the book *Alive!* by Piers Paul Read. Ironically, the survivors could have avoided such hardships if they had chosen a different escape route from the crash site. They had gone north into impossible country, instead of south, where they could have easily picked up a trail leading them

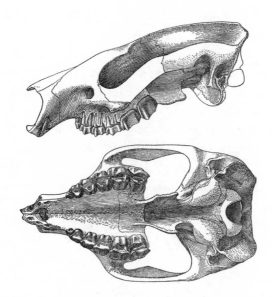

The skull of a notostylopid, a fossil mammal from the Early Tertiary of South America, well represented at Termas del Flaco

to Termas del Flaco, where the resort, though boarded up in winter, had cans of food that could have sustained them. I thought how easy it was to make that judgment from our high vantage point and how disorienting and difficult it was for those unfortunate, valiant Uruguayans.

In the morning, the streams around camp hissed softly before the snowfields that fed them were warmed. The peaks were deep in black shadow against the rising eastern sun. By late in the long afternoon, the throne of marine limestone that towered over us became a dazzling gold icon. Clouds formed and condors sailed in great circles at our feet. Like guardian beasts, the giant birds perched in clusters on the rocky cliffs containing our brilliant fossils. When the spell of good weather was broken by a mountain storm, the result was intimidating but magnificent. Horizontal lightning sparked out of clouds drifting into the deep Tinguiririca Valley from the Pacific. One storm was powerful enough to wreck our tents and flood our camp. We were concealed by swirling clouds for a day and a night from the town far below, where the staff at the Hotel Gloria lit votive candles and prayed for our safe return.

The discoveries above the Tinguiririca Valley launched over a decade of paleontological and geological work that continues to this day. Soon it became apparent that the fossil animals at the site were a unique combination of species, and they seemed to represent a time interval in the late Eocene not represented elsewhere on the continent. Supporting this important discovery was a set of radiometric dates taken on the rocks that entombed the fossils. It was our good fortune to have fossil beds actually interleaved with beds of lava that could be sampled for radioactive minerals. Analysis of these minerals gave a date that made perfect sense: about 35 million years in age. We could not only identify a new time interval—which we called the Tinguiririan age—based on the fossils, but also provide it with a good bracket of dates ticked off in millions of years.

This set of findings was capped by a very significant discovery. At a Miocene-aged locality in the region, the team recovered a beautiful jaw of a primitive New World monkey. The discovery, along with the rest of the animals from the Andean main range, attracted worldwide attention.

The Andean Paleontological Expeditions gathered momentum through the years, greatly strengthened by the collaboration of Renaldo Charrier, an excellent geologist from the University of Chile in Santiago. The cooperative effort was eventually carried out in a nation that changed for the better. A million people stood in the streets of Santiago in January 1990 to rejoice in the deposing of Pinochet and the beginning of a new democracy. By the first years of the decade, John Flynn and Andy Wyss, now in charge of the expeditions, struck out in all directions scouring the vertical ridges that separated one deep Andean river valley from another. They complemented traditional and erratic travel on horse with the convenience of helicopter flights. What was once an uncertain reconnaissance in one of the highest and most unfamiliar pieces of crust on earth had become a series of well-funded and outrageously productive international expeditions. The results have been adding an important slice of time and set of organisms and environments to the evolutionary history of South America.

As for me, I enthusiastically passed the leadership of the expeditions to my comrades. In 1991 I played a delightful role in assisting my former student, following Andy's order to prospect a high set of beds farther down the Tinguiririca Valley. There Mark Norell and I forded a wide but fortunately shallow part of the river and climbed the cliffs to a new layer of fossils. We then made a three-day traverse at 11,000 feet, prospecting the producing layer as we moved eastward to reach our original camp perched high above Termas. En route, we wandered into waterfalls several hundred feet high, magnificent slopes of crystalline snow, and lichen-filled grottos dripping beads of water that shone like diamonds. It was easy to imagine that no one had ever seen these places before. But during that 1991 season I knew I was already completing my own work in those splendid mountains. The wonders of Asia and its astounding fossils had already captured me.

22

. . .

The Land of Sheba

"Meester Ian, Meester Mike, these are veery bad men, they will keel us," was Ali's warning call. Ali was our sentry, our Yemeni escort from the Ministry of Oil and Mineral Resources.

As I scrambled out from my tiny bivouac tent, I could see his trembling figure silhouetted in the glare of headlights from an alien vehicle. Then I saw several heavily armed men jump out of the jeep and come stalking toward us. One surly fellow pointed his Soviet-built automatic rifle at my forehead. A sullen leader stepped forward and started speaking.

In Yemen, as in other countries on the Arabian Peninsula, speech is often notably emotional, forceful, and deceptive. To those familiar with neither the culture nor the language, a casual greeting and an exchange of news can sound like an argument, and an argument can sound like a declaration of war. But there was no mistaking, in this case, the commander's message or motive. He would have his way or slit our throats.

Poor Ali. He was so terrorized he could hardly speak, let alone translate. Jeff Meissner, director of the American Institute of Yemeni Studies (AIYS) and our host, took up the slack and started speaking in very polite tones to the truculent leader. Then he turned to me and my colleague, Ian Tattersall, and composed a translation of the fellow's message—very carefully, lest one of the band by some small chance understand English.

"This officer commands a patrol of the North Yemen border militia. He says that this is a very dangerous area—full of anarchists, gunrunners, smugglers, and hijackers. We are trespassing without

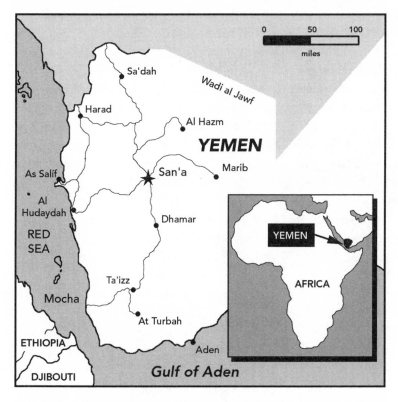

Yemen

any official permission or purpose. He forcefully 'invites' us to sleep at his army checkpoint."

We had already spent most of the hot month of June 1988 exploring North Yemen, and we were thoroughly acquainted with the intrigue and duplicity of such a message. By now it was June 29 and we had only a few days left to complete our work. Was this really a government militia patrol, or simply one of the myriad autonomous bands of armed men in khaki? Did this band represent official business or anarchy? Were these men really protecting us or were they threatening one of the very dangers—hijacking, robbery, or even murder—that they were so earnest in condemning?

We had been taught to assume the worst in any encounter, and then to be delightfully relieved if the situation suddenly improved. In this case, however, there was no clear reason for even cautious optimism. The commander barked an order, and the intimidated Ali hastily produced his official identity card.

"Big mistake," Jeff murmured in a tone that suggested he was actually declaring, "That man is dead."

The commander now gave us an order we could all understand without translation. We dutifully dismantled the tents and packed our Toyota Land Cruiser. Unfortunately, Ian had some bad news. "I can't find the keys," he declared as he dropped to the ground and began crawling around the vehicle. We joined in the search, a spectacle that did not please the commander. He yelled more orders and the soldiers lunged toward us with their guns.

Ian suddenly stood up. "Here they are!" He extracted the keys from one of the twelve pockets on his outdoorsman vest. We immediately jumped into our Land Cruiser and started down the winding dirt road on a mountain slope covered with boulders. The militia jeep tailgated us, ready to give the signal to turn at the proper junction toward their mysterious encampment.

The situation was all the more frustrating because we seemed to have been sitting ducks for capture. Only a few hours earlier we had come along this mountain road and crossed a makeshift checkpoint, where an old guard blithely allowed us entry. We had then negotiated its steep switchbacks heading toward the summit of the 10,000-foot Jebel Marah and eventually found our pleasant campsite, a high mountain pass of promising sandstone sprinkled with junipers, with a grand view of the mountain country extending north to the Saudi Arabian border. At Ali's urging we had even dutifully returned to the guard post and reported our intentions to the old sentry, who, after hesitating for a moment, had permitted us to camp. Now it was clear that our prudent report to the "authorities" had merely given this so-called commander and his patrol good information on our whereabouts and made it easier to apprehend us.

Ian tried to get some useful information from our quaking escort, now stripped of his documentation and official identity.

"So, Ali, what's our gracious commander up to?"

"He is a very bad man," Ali replied without offering any more illumination.

Yemen was a strange land of contrasts; a stunning, wild landscape formed a backdrop for some of the most sublime architectural wonders ever created, but these glorious cultural contributions emanated from a society accustomed to frank and sometimes barbaric brutality.

Much of this violence was carried out in keeping with an ancient code of honor, morality, religion, and cultural patrimony. It would be hypocritical to claim that Yemen was different from all other cultures or countries in its use of such a code, yet it did seem to have a particular style of and emphasis on ferocity. Since our arrival in Yemen at the end of May, we had been told of a tableau of violence either sanctioned or simply ignored by Yemeni society. Depending on the tribe or the situation, execution or dismemberment was a broadly applied remedy not only for serious crimes but for petty thievery, adultery, and lesser transgressions. One execution method, we heard, was allegedly death by explosion, wherein the condemned was filled with air from a compressor-driven tire pump. We knew that a group of foreign geologists in a Land Cruiser like ours had been approached in the desert by a small band of tribesmen. Their Yemeni escort, from the capital city of San'a, who had the misfortune of being associated with a tribe despised by the locals, was pulled out of the vehicle and executed, though the geologists were invited to explore and enjoy the region.

In other cases, foreigners were less fortunate. There were reports of killing and kidnapping. Women were especially vulnerable, and there were verified tales of women being imprisoned and tortured in some black citadel perched on an isolated crag in the mountains. Less than fifty miles outside the city limits of San'a, every man from sixteen to seventy was armed with an automatic rifle and a *jambiah*, a wicked curved knife with a handle made from the horn of an endangered species of rhinoceros.

Even the capital city was not insulated from the ambience of violence. The most popular cultural spot in town was the war museum, where proud fathers guided their sons and future warlords through halls and courtyards displaying piles of arms, mural-sized photos of decapitated and dismembered corpses, and a bullet-riddled, bloodstained limousine that had served as the royal carriage of an assassinated imam. A short time before our arrival, we were told, the American Consulate was hit with a missile, which fortunately missed the party on the patio and exploded in an unoccupied bathroom on the second floor. The official government response in San'a to this mischief was a desultory investigation that concluded with the dubious explanation that the perpetrator was "an irate Peace Corps worker." Worse, in 1988 North Yemen, or the Yemen Arab Republic (YAR), was still in conflict with the communist South Yemen, the People's Democratic Yemen Republic, over a disputed border. The YAR was also a declared ally of Iraq's Saddam Hussein in his current war with Iran. American Museum officials had advised caution when Ian and I announced we were heading off for the land of the Queen of Sheba. As we had when we went to Chile, we were required to carry an extra policy of "war zone insurance" before proceeding with the expedition.

The obvious question was, why Yemen? Why not do paleontology in more auspicious and tranquil territory? I certainly reviewed those questions as we turned off the main road to the camp— and to our incarceration. We knew, however, that Yemen offered dramatic possibilities for fossils. A huge expanse of the country, especially in the northern region, was covered with Jurassic-aged limestones and shales called the Amran Series. This unit contained an array of small marine creatures, but there were also reports of large bones—possibly dinosaur bones. And lying above the Amran were the alluring Cretaceous Tawilah sandstones, whose continental facies might also contain dinosaur and mammal bone.

The younger rocks of Yemen were also intriguing. Until about five million years ago the Arabian Peninsula was still attached to the African continent. Tens of millions of years before that final break, the Red Sea had started opening along a tectonic plate boundary as Arabia rotated counterclockwise away from Africa. The gash between these two huge landmasses marked by the Red Sea is actually the northern extension of the Rift Zone in East Africa, a great complex of faults on a scale unmatched on any other continent. The Red Sea had originally opened into the Mediterranean, and with this breach, a northern land connection for the dispersal of species between Africa and Asia had been severed. Over the next few million years the sea unzipped in a southerly direction, eventually making its incursion between the heel of Arabia and the horn of Africa into the Indian Ocean. As Peter Whybrow, a paleontologist from the Natural History Museum in London, argued, the only such land connection between 30 and 5 million years ago had to be right through Yemen. Thus, Yemen and the surrounding regions of the western Arabian Peninsula might show an interesting mosaic of fauna, a mix of animals well represented in eastern Africa and in southwestern Asia, in areas like Pakistan. Perhaps these animals might include human ancestors, as hominoids were preserved along with ancient elephants, large cats, and other carnivores, horses, and giraffes in the important Miocene rock sequences exposed in Kenya, Tanzania, Uganda, and Ethiopia; contemporaneous faunas in Asia were similar but not identical to these. Moreover, there was a tantalizing similarity between the huge Miocene rock sequences in places like Ethiopia that had produced fossils and the unprospected rock layers of Yemen. A good batch of fossils from Yemen might fill in the gap along the great corridor of exchange between the two continents. Owing to its close proximity and late connection with Africa, Yemen might also yield fossils representing the first appearance of hominids in Asia.

The hunch that Yemen was a possible crossroads for the dispersal of hominoids, other mammals, and other vertebrates appealed to Ian

Tattersall, an anthropologist colleague at the American Museum of Natural History. Ian raised money from the National Geographic Society and invited me to join him in the summer of 1988 for an extensive survey of Yemen. Ian is a multilingual, highly cultured child of the former British Empire. He was raised in Kenya, trained at Yale, and has a powerful expertise that ranges from science to language, fine cuisine, and wine. In the course of his studies of hominid and primate paleontology and biology, he had made a mark with many publications. Ian even had the distinction—not a trivial one—of having the rare, elegant lemur *Propithecus tattersalli* named after him by another scientist. Notwithstanding his tastes for high culture, Ian was a roamer, fully inured to the uncertainties, discomforts, and dangers of traveling the forgotten corners of the world.

Despite the enthusiasm we shared for the Yemen adventure, Ian and I recognized that the odds against achievement were high. Information on the country's geology was literally remote: the rock units of Yemen had been mapped primarily from satellite photos. Ian strategically chose a third member of our expedition to make up for this deficit: Maurice Grolier, a retired scientist from the U.S. Geological Survey who had made the geologic map of Yemen from Landsat images and had spent some time ground-truthing his map by roaming the Arabian countryside. This was not Maurice's only close encounter with space. His specialization in desert geology qualified him to offer a choice for the first manned lunar landing site. He picked the gargantuan, smooth-skinned depression, the Sea of Tranquillity, and won. "I merely picked the smoothest spot," he admitted.

Maurice seemed more annoyed than agitated by this apparent Yemen border patrol. During World War II, he had been captured by the Nazis and spent two years as a prisoner of war. As a civilian, he had experienced many bizarre and threatening events in his wanderings in Egypt, Sudan, and the Arabian Peninsula. It would take more than a confrontation with a motley militia to intimidate him.

"These imbeciles, they are trying to be important," he scowled. When we reached the patrol's camp, a scatter of limp canvas tents near a dilapidated tank stranded in a dung field, Maurice's bulky frame eased out of the Land Cruiser with no more urgency than he had on any other field stop.

But the commander now displayed enough dissatisfaction to remind even Maurice of our predicament. During the ride he had apparently become extremely incensed about our refusal to hand over our passports with the same cooperative ease that Ali had offered up his precious identity card. He started yelling again, demanding our original passports.

Now even Jeff was riled. "If we give him those, we've had it," he instructed. Jeff had indeed done the right thing and left his original locked in the safe back at the AIYS in San'a.

We produced only Xerox copies of our passports and our official carte blanche from the Ministry of Oil and Mineral Resources. The commander insisted again on our originals. To this, we played dumb, and started to unpack our gear, feigning an overwhelming need for sleep. I asked permission to urinate, and walked around the corner of a small pile of boulders, momentarily out of view of the soldiers. Here I quickly buried my original passport under a bush.

Meanwhile the soldiers demanded the keys of our Land Cruiser and parked it some yards away from our enforced bivouac. I envisioned an unfortunate scenario where the Toyota was ripped apart piece by piece in search of original documents, money, and other items of value. Then something wholly unexpected happened. As the commander bid us almost a cordial goodnight, the soldiers returned the keys to Ian. Dispelling our momentary relief, Ali then informed us that the commander would not let us go the next day. There were some vague procedures that had to be carried out—more interrogation, a careful scrutiny of the Ministry letter (we suspected the commander was not very literate), and a phone call to San'a. It was not at all clear to us that these were the real reasons for our prolonged detention.

The commander and his royal guard finally hopped into their jeep and sped up the hill to a main camp, leaving us to a sentry detachment of young boys with AK-47s. The conditions were not wholly conducive to sleep, and I lay there staring at the blackness of the Arabian night as a heavily armed teenager trudged back and forth only inches from my head. I could tell the others too were restless and awake, except of course Maurice, who soon subjected us to a thunderous snoring.

At last I lost consciousness, dreaming of imbroglios over official papers while people from New York drifted in and out of a scene that seemed to be set in a mosque. There was an anxious struggle with a guard on a parapet, and the pitch of the battle suddenly awakened me. It was 5:00 a.m. and as cold as any dry desert can get after the heat of the day has been released into the nocturnal atmosphere. Wrapped only in a thin cotton blanket, I trembled with fever and a cold sweat. As the young soldiers drifted aimlessly about our sleeping area, I waited for a warm sun and contemplated the dismal state of affairs. Those predawn hours offered time to reflect on our adventures over the past several weeks.

We had arrived in San'a on June 1, in the veil of an early morning much like this. The capital city sprawls over what seems to be a stark plain of tan and burnt sienna, but is actually the caldera of an enormous extinct volcano, set at 9,000 feet above sea level. I was surprised by the coolness of the San'a dawn and shocked by the volume of the morning prayer issued through scratchy loudspeakers that crowned hundreds of minarets. Jeff showed us to our rooms in the AIYS, a magnificent mud brick building with stained-glass windows and a courtyard full of fruit trees and grapevines.

By the next day we had already accomplished much official business with His Excellency Ali Jabr Alawi, Deputy Minister of Oil and Mineral Resources. We even forged an agreement to make a fifty-fifty split of any collections of fossils recovered. This was great news, as such permission from foreign countries was becoming increasingly difficult. "Now all we have to do is find the bloody fossils," Ian declared.

To this end we scoured what seemed to be every ravine, cliff face, depression, and road cut in this rugged, incredibly beautiful country. Our first jaunt took us down the great escarpment west of San'a that straddles sharp-edged peaks more than 10,000 feet in elevation, and to the Tihama, the scorching plain that fringes the Red Sea. En route we passed Tawilah, a town perched at 7,000 feet on a prominence that is the namesake of the spectacular Cretaceous Tawilah Sandstone that dominates the surrounding landscape.

I had never seen architecture so elegantly sutured to geology. Tawilah the town was merely a more ordered outcrop of the stunning beige- and rose-colored sandstones that embraced it. In other places, the jagged black shoulder of a hulking volcano in the distance soon resolved itself into a town of multistoried buildings forged from the surrounding lava. Then there were the northern villages, cut like cliff dwellings in the swirling crossbeds of the Permian Wajid Sandstone. In the southern reaches of the country, near Ta'izz, buildings were magnificent facades of limestone bricks in all shades of grays, greens, and subtle blues quarried from the terraced limestone mountains around them. Where there were no nearby rock formations of any magnitude, mud would do. Not only major cities like San'a and Sa'dah, but also the villages on the fringes of the Empty Quarter—or Rub' al Khali, the sandiest desert on earth—were constructed from mud brick. These mud towers were either drab and uniform in color or lavishly decorated. The ancient center of San'a, a city of more than 400,000, was a fantasy of mud brick buildings boldly decorated with lattices of brilliant white that bordered windows, doors, archways, and parapets. In parts of the desert, the mud buildings were painted in aggressive patterns of red, blue, and green—like giant Navajo rugs. Other towns simply left their mud castles unadorned and relied on the symmetry and elegant profile of the structures to enchant. Yemen was one of the rare places on earth where the extraordinary mastery of architecture actually drew one closer to the monumental grandeur of the natural rock formations.

In addition to the marriage of building and mountaintop, Yemen had other marvelous features. The spice markets were laid out with piles of cumin, fenugreek, cardamom, coriander, cloves, and turmeric, enough to delight any ancient king making his way along the frankincense trail to the Holy Land. On the shores of the Red Sea we savored fish blackened in tandoor ovens and slathered with tomatoes, chiles, onions, garlic, and fresh cilantro, a concoction that verged on being as good as the best salsas from Baja California. In

A Yemen house

San'a, women virtually covered in black but wearing brilliant red
high heels approached you with frank and alluring stares from eyes
that shone like fire opals. In the vernal south around the city of
Ta'izz—where Muslim fundamentalism was relaxed in favor of
much older cultural traditions—the women revealed their beautiful
faces and wrapped themselves in shimmering silk tapestries of
green, silver, and red, as they balanced heavy water jugs or bundles
of wool on their black hair. We were often visited in camp by local
countrymen who brought us honey-laced tea, fresh milk, and
bread, and occasionally a delicious disk of smoked goat cheese.
Some of our camps were planted in the most beautiful oases imaginable—a putting green of grass surrounded by date palms where a
flock of chromatic weaverbirds made their nest. At a scenic overlook at the edge of one of Yemen's main highways, I stood in awe
and scanned the decorated towers, walls, and intricate arches of the
town of Kawkaban, rising like the ringed fingers of an Arabian
beauty from the wind-sculpted walls of sandstone at its footings. A

Yemeni man approached me and asked in halting English, "You like?"

"Yes, yes, more glorious than all the castles of Spain," I answered.

The man grimaced, almost insulted by my remark. Then he replied, "Spain? I do not know Spain. There is no reason to leave this country."

It was a common response. The Yemeni were fiercely proud and superior about their people and their land. They had a profound reason to be so.

But all this wondrous elegance was blemished by some unfortunate accoutrements of civilization. These less attractive features of Yemeni life were not confined to weapons and expressions of violence. Plastic was a technological wonder gone mad in this country. Every article of clothing, food, and medicine we received was dispensed to us in a plastic bag. Unfortunately, there seemed to be no provision for recycling these bags or even burying them. Every bush around every village seemed to bear fruit in the form of blue and pink plastic bags placed there by the swirling winds. Sanitation in some areas was another problem. In the town of Hodeida, on the banks of the Red Sea, we checked into a reasonably respectable hotel where one was forced to shower in a basin that did double service as a toilet. After sweating profusely in the heat and humidity of the Tihama it was difficult to contend with a shower floor ornamented with piles of human excrement. There were also those surly sentries and their proliferating inspection posts. It was common to drive along a deserted and unguarded road and then, on the return trip, to find a new checkpoint—or two, or three— miraculously sprouted in the middle of nowhere. On our first drive in early June down to the Tihama we rejoined the main highway via a ragged dirt road at a junction completely neglected by road crews but warranting a military checkpoint. It was dark, and I was delirious from the ten-hour drive required to cover a mere 160 miles of backcountry road. As I drove to the left of an upright oil drum painted with the red, white, and black of the YAR flag, two infuriated soldiers started yelling at me. Apparently I had driven on the wrong side of the drum.

Other aspects of Yemen life offered an ambiguous mix of benefits and liabilities. Perhaps most prominent among these was *qat*, the
shrub with narcotic foliage like bay leaves, chewed daily by virtually all the Yemen Republic's four million men and, more surreptitiously, a smaller contingent of women. *Qat* was rather pricey, and it
drove a *qat* economy, made terracing and farming profitable, and
obviously brought pleasure to those who chewed it. But *qat* had a
stranglehold on Yemen. In the early afternoon every activity came
to a halt as the workforce was lulled into a soporific state by the big
chew. The main highways, the asphalt green with the stains of *qat*,
were lined with big trucks parked on the side, which provided
shadow for listless drivers. The profit motive of *qat* had driven a
more sustainable agriculture to near-extinction. The terraced hills
of Yemen were a significant portion of the precious few bits of
land on the entire Arabian Peninsula that were good for a variety of
crops—coffee, bananas, dates, grapes, and other fruits and vegetables. These once richly diverse slopes had been transformed into
monoculture plots for *qat* growers, where soil gave out too fast
without the benefit of crop rotation. Much of Yemen's terrace
country looked like the weed-choked ledges of an ancient Assyrian
step pyramid.

Our travel routes through this land of volcanoes, enchanted castles, sand deserts, and inevitable *qat* fields involved extended sorties—loop treks that took us out to some remote corner and in a
few days brought us back to San'a to recuperate and fill out more
papers for the next jaunt. On trip number one, when we reached
the Tihama we headed north through arresting scenery; the Tihama
looked like an African savanna that stretched from the mountains
to the sea. The flat panorama of yellow grass was interspersed with
elegant shade trees known as *teluga*, which reminded me of the live
oaks in the once-pristine San Fernando Valley north of Los Angeles. But we were disappointed by the geology of the Tihama. The
land was dissected with only low exposures of friable sands, too
weathered to contain good fossil bone. We ended our northward
journey at the unattractive, dusty, junk-ridden town of Harad, near
the Arabian border. Harad had only one asset. As we had the engine compartment of our Land Cruiser sucked free of a mountain

of red dust, I staggered in the heat over to a Coke machine and managed to get it to cough out a semicold bottle of the elixir. Nearby, two *qat* chewers were lying on the coils of old bed frames in the shade of a *teluga* tree, stupefied. A television placed outdoors on an old cabinet, its picture outshone by the blinding sun, crackled valiantly with a strange Japanese cartoon overdubbed in Arabic.

In the region of Harad we scrambled over outcrops of the Amran Series, looking for dinosaurs, and Baid Formation sandstones, looking for Miocene mammals. Neither quest was successful. To make matters worse, the geologic maps—which Maurice admitted were made with a combination of space and aerial images, minimal ground-truthing, and much speculation—failed to match much of what we could see with this more intensive scouting. An extended outcrop of sandstone on the map was in reality an unimpressive road cut or trench. Some of the smaller outcrops indicated on the map didn't exist at all. Maurice found the whole operation very enlightening insofar as his efforts to improve our geologic knowledge of Yemen were concerned, but this was little consolation for a bunch of desperate bone hunters.

From Harad we drove southwest toward the shore of the Red Sea, the great ancient gash between two continents. At dusk we passed the sleepy village of Al Luhayyah. With its odd cylindrical huts and its domed-shaped, thatched roofs, the town looked more Ethiopian than Arabian. A few of the domes were eerily crowned with TV antennas. At As Salīf (*salīf* means salt) we pitched camp and spent a fitful night assaulted by the buzz of mosquitoes and the howling of dogs in a night that felt hotter and more oppressively humid than the day. The next day we prospected gypsum-laced sheets of rocks barren of fossils. Then we turned back east and took the long and winding road through the mountains, reaching San'a by nightfall.

Travel from the seacoast to the high plateau of the interior wasn't always so easy. On December 29, 1763, a famous Danish expedition led by the redoubtable Carsten Niebuhr landed in the harbor of Al Luhayyah. Over several months the team, which comprised an Arabist, a botanist, a physician, and an artist, wandered the Tihama down to the coffee port of Mocha and then slowly strug-

gled up the great escarpment toward San'a. But the group was drastically cut down by infectious diseases. When the scientists reached
the plateau they encountered pockets of both the hospitality and
the xenophobia that still characterize the country. Instead of receiving welcome and sustenance, they were nearly stoned to death at
one village. Eventually all members of the expedition died save
Niebuhr, who returned alone to Denmark in 1767 bearing exotic
tales of both the pleasures and the hazards of travel in Arabia Felix.
Niebuhr's famous observations included the news that drinking tea
from husks was more common than drinking coffee made from
beans, qat was unfavorable to sleep, Europeans were not allowed to
ride donkeys in Mocha, the streets of Jiblah were paved, Yemeni
girls marry as early as nine or ten, and the hills around San'a were
already deforested.

Disease is still a serious factor in Yemen exploration. We had prepared with the usual inoculations—tetanus, typhoid, yellow fever—
as well as prophylaxis in the form of powerful lariam pills to protect
against predictable strains of malaria. We had overlooked one important preparation, however. Yemen was full of dogs, most of them
noisy and rather aggressive. On certain occasions, I would open the
netting of my bivouac tent in the early morning to see it guarded
by a strangely passive cur that rolled its eyes and foamed at the
mouth. No one should travel in the backcountry of Yemen without a rabies vaccination.

The trip to the Tihama and the Red Sea was an uncomfortable
paleontological failure. After a few days of fruitless scouting of the
Tawilah Formation near San'a, we next turned our attention to the
volcano country around Dhamār some seventy miles south. Again
the venture was without reward. The highlights of this scouting trip
were magnificent lava-sculpted villages characterized by laughing
children, packs of barking dogs, and rumors of intertribal feuds and
warfare. For much of June 14 we pioneered a very uncertain route
through twisting gullies at the base of a huge volcano, only to end
up at our starting place. We ended this folly of a journey with an
unpleasant night in an ugly modern cluster of concrete block
houses. These edifices were built by a Saudi organization for protection against earthquakes, but they were absurdly inferior to the

style and grace of the traditional, albeit less earthquake-proof, Yemeni dwelling. The housing tract was virtually deserted by people; only packs of feral dogs roamed their own private stronghold with impunity.

By June 15 we were at last finding fossils—but clearly of the wrong kind. Some very young Pleistocene rocks near a road that crossed the southern part of the escarpment were chock-full of clams, snails, and other invertebrates, an assemblage that neither surprised nor excited anyone. "Maybe there are fossils of humans who ate this stuff," I said. We looked in vain for such a remote possibility. The next day we had a surge of success when I found a chunk of fossil bone—real bone—weathering out of a cliff flanking the Hodeida–San'a highway. At closer inspection, however, the fossil looked very young, probably only a few thousand years young. *"Merde!"* Maurice sneered as I pronounced my verdict on the dismal value of my prize.

After recuperation and bureaucratic maneuvers in San'a we headed out once again on June 19, due south to the high hill country around Ta'izz. Near radiant flamboyants, or "flaming trees," we prospected ancient lake deposits finely layered into clays and silts. These very auspicious rocks still offered no fossils worthy of our attention. Much the same beds produced plant fragments, but no bones, over the next few days. We ended this excursion in the craggy country around the village of Ar Rawnah, an allegedly dangerous area full of feuding tribes and insurrectionists. In a foreshadowing of our later detention in the north, I was once awakened by the poke of a gun muzzle on my tent netting. Our camp in a beatific oasis had been invaded by a group of makeshift militia looking for Marxist guerrillas and smugglers. Determining that we were neither, these soldiers turned out to be quite affable. We sat with the patrol in the morning shade of a stately date palm aside the ravaged mountain road, sharing their hot tea and bread. A withered old man came along beating an equally withered camel with a pale, near-albino coat of patchy fur. The old man looked at us, contorted his wrinkled face, and started screaming something in regional dialect that seemed as incomprehensible to the soldiers as to us. Then he went on, beating the flanks of his camel as it reluctantly climbed the ridge above the oasis. The soldiers laughed at the spectacle. One of them turned to us and said something that sounded like an Arabic proverb, at which the others roared. Jeff smiled and translated, "He says the old man wants his lover all to himself." Apparently, rude jokes about camels and human bestiality were also part of the local scene.

The Ta'izz–Rawnah excursion was perhaps more scenic than earlier outings but no more productive. Back in San'a we realized that time before our early July departure was slipping away. I had obligations to travel back to Wyoming for fieldwork that summer and later in August to Sweden for a Nobel conference on evolution. Ian was equally well booked. It was already June 25 and we had spent three weeks of intensive field checking without anything to induce us ever to return to Yemen, at least for its paleontology. I tried to recall a more unsatisfying outcome for all our effort and time, but could not. The dull weight of defeat was somewhat lightened by our meeting with a professor at the San'a University, Dr. Hamed El-Nakhal. The professor, whose simple last name translates to a much more complex phrase—"he who makes a sieve out of palm fronds"—had mined Jurassic-aged fish in a gypsum quarry north of San'a near the town of Al Ghiras. But Hamed was an invertebrate paleontologist and stratigrapher and hadn't spent much time looking for bone. We resolved to visit the quarry with Hamed at the end of the trip after we had exhausted all other possibilities.

Those remaining prospects meant travel to two areas that we had thus far avoided: the far northern section of the highlands and, to the northeast, the great Arabian Desert itself. There were reasons for this hesitation. Both regions were even more poorly mapped than the others, and both had more bandits, aggressors, smugglers, and reports of violence than areas we had already scoured. But Ian and I had not found a single good fossil of interest; we were determined to go against the advice of some of the local officials and foreign oil geologists in San'a and explore these less welcoming locales. These forays also required an excruciating wait for our passports to be approved with official extension beyond the terms of our tourist visas. At last, on June 28, we received our newly stamped passports and were ready to head north to Sa'dah and the smuggler country beyond. During the previous weeks we were accompanied by a rotation of affable and well-trained scientists from the Ministry of Oil and Mineral Resources. This time, our Ministry escort, poor Ali Mohammed, drew the short straw and was required to accompany us. Thus on the night of June 29 he had the misfortune, along with the rest of us, of sleeping in a dung field guarded by teenage gunmen.

The next morning, sunken eyes and ashen faces indicated that my colleagues—excluding Maurice, of course—had slept as poorly as I had. We sipped tea under the eyes of the juvenile sentries and a sun rising above the eastern ridges of Amran limestones. At 8:00 a.m. we saw a cloud of dust rising over the lip of the hill that concealed the main camp. Within minutes a small military truck came racing toward us at a suicidal speed. The truck pulled up and an old friend—the same rascal at the guard post who had originally allowed us entry to this forbidden zone—jumped out and came over, looking rather downcast. To our great surprise he reached into his pouch and handed over our passport copies and Ali's ID card. He then announced we were free to go but forbidden to return to this "high security" zone.

So what had happened? Why this rather undramatic resolution to events that originally seemed so dark and threatening? Maybe the sullen commander finally got a reading of our impressive letter from the Ministry. Maybe he was castigated by someone of higher rank. Certainly the guard who was the original source of our troubles had received a severe reprimand, for he was now sheepish and subdued. Perhaps it had sunk in that our Ministry letter was the real McCoy and meant big trouble to anyone who put us in harm's way. It seemed likely that this militia concluded that as long as we were banished from their realm they could go on with their usual business, whatever that might be. We shall never know. We were mystified but highly relieved. After all, in more recent years such arrests or kidnappings of foreigners with more fatal outcome have been part of the news coming out of Yemen. At any rate, I cannot recall ever feeling less regret at being denied permission for prospecting a piece of land for fossils.

I drove down the Jebel Marah road as fast as that military truck had, just in case the decision to release us might be reversed. We all kept looking back up the road for the commander's jeep. Some twelve miles away and back on the main road we pulled over, relaxed enough to make some tea and repack the truck.

"We can be back in San'a before the night, *Inshalla!*" Ali announced with joy.

Ian and I merely smiled at each other. Then Ian gave Ali the bad news.

"Ali, I'm afraid it's not homeward yet. Mike and I want to go to Wadi al Jawf."

Ali looked at us in horror. "Oh no, Meester Ian, there are veery bad men in the Jawf, they will keel us."

I warmed my arms in the morning sun as I squatted cross-legged like a Bedouin warrior on a toadstool of Wajid sandstone, and thought about Ali's panic attack. Our Ali Mohammed was no Muhammad Ali, although there was some reason for his agitation. Wadi al Jawf, a great wash that drains its sands into the Empty Quarter of the Arabian Desert, is perhaps the most isolated and autonomous region in Yemen. It is an area full of defiant and sometimes hostile tribes, brutal frontier law, hijacking, and most of all, uncertainty. Yet Ian and I were desperate for fossils, and this was one of the last places to look, despite our very fresh experience with threats and detention. Ali must have thought we were insane. Even Jeff was subdued in response to our decision. He had earlier voiced reluctance to take the only AIYS vehicle to the Jawf. He could only comfort himself with the notion that the vehicle "was worth more to them than we were."

We returned to Sa'dah and then drove on to Al Harf, a small village and truck stop where we turned off the highway and headed east on a series of braided roads that intersected with the northern reaches of the Jawf. Four miles from town the Land Cruiser slumped as its left rear tire exploded. As we jacked up the car and replaced the tire with a mangy spare, Ali was exultant, assuming this meant a return to San'a. To his dismay we returned instead to Al Harf, managed to fix the tire for 30 rials (about $20), and resumed our eastward tack to the wild and forbidding desert.

Twenty-four miles of driving brought us to the vastness of the

Jawf, a series of broad washes flanked on the horizon by low cliffs of Precambrian weathered granite topped by Amran limestones. Anxiously we scanned the cliffs for any sign of rock layers that might be younger than the Jurassic, but could find nothing. In the late afternoon we reached the town of Al Marāshī, a cluster of beautiful mud buildings boldly decorated with red-and-white patterns that framed doors and windows. The whole scene was disarmingly peaceful and unthreatening. Throngs of children danced around our car as parents smiled approvingly from the doorway of their stately mansions. Even the young women smiled and posed for Polaroids.

We reluctantly departed this serene village and made our way farther southeast, moving along a plateau above the main wash. Here we inspected some low hills that from a distance looked like tantalizing Tertiary rocks but up close again proved to be Jurassic Amran limestones and shales. It was Amran everywhere. I almost prayed, offering up a thousand rials for any fragment of dinosaur bone I might find in that ungenerous sediment. As dusk approached, we stopped bashing our heads against the barren rocks and looked for a decent and reasonably isolated campsite. We found one in a tributary drainage called Wadi Dayr, at a graceful bend of a sand river fringed with acacia trees. Ali became more agitated and reminded us of the consequences of our last camping outing. We were all, in truth, on the alert. But it was difficult to remain edgy; Wadi Dayr was so stunning. I slept in the open on the soft white sand bed of a wash that glistened like an ice field under the full moon. The acacia branches made millions of graceful reticulations against the lunar light. It was the desert of poetry—of Omar Khayyám, the European waif Isabelle Eberhardt, Paul Bowles, and Muhammad himself. I dreamed of ancient wells, date palms, and mysterious, beautiful women with henna-stained hands.

The morning in Wadi Dayr was as enchanting as the previous night. A fine mist, not as substantial as a fog, scattered the light and wrapped the acacias in diaphanous shrouds. As we packed the truck three men with guns slowly approached us through the mist. I thought not only of a potential threat but of Ian's observation about many parts of Arabia, Africa, and the developing world: that despite the sweep of emptiness and solitude in these places, one was never really alone. People would suddenly materialize behind a bush while you were making a latrine squat. The world, even the so-called uninhabitable world, was full of people. To our collective relief, our visitors threw their automatics to the ground and simply sat, cross-legged, to watch our routine. Ali placed himself at some safe distance from the men but Jeff struck up a conversation with the oldest, a small lean fellow of about thirty. He informed Jeff that he and his cousins were the *thum'n* (eighth) of the Al Shamma of the Dhu Nasayn, part of a complicated hierarchy of one of the great tribes of the Jawf.

More conviviality down the road at the hot springs of Hamman al Wagarah continued to dispel our apprehensions about the putative terrors of the Jawf. At the edge of the gurgling spring young women laughed and combed their sisters' wet hair. Men gathered around our vehicle making friendly gestures and carefully examining every door handle, every patch on our road-beaten tires. For the benefit of my camera, an old fellow with barely half his teeth playfully grabbed one of his comrades in a head lock and held his wicked, curved *jambiah*, his warrior's dagger, against the other's throat. They both howled with laughter at their own theatrics. An occasional breeze shook the acacias and cooled the sweat on my back. Unlike so many other places in Yemen, there was no evidence here of official military control over access to a valuable spring or oasis. The area was also pleasantly devoid of other signs of government constraint—checkpoints with painted oil drums and propaganda posters of the dictatorial Yemeni president, Ali Abdullah Sali. The Jawf seemed left alone to the desert and its good-humored inhabitants. I began to like the place very much.

When later in the day we finally did encounter a checkpoint, the men stationed there seemed more intimidated by the vastness of the Jawf than surly or threatening. With barely a glance at our passport copies they bid us on our way. The next encounter was slightly more aggressive. A cluster of men in a Land Cruiser that bolt for bolt matched our own sped up to our flank and commanded us to stop. The men got out and craned their necks inside our open windows, surveying our compartment. A brief exchange between Jeff and the obvious leader ended with a nervous laugh from Jeff. Our visitors then jumped back into their Toyota and mysteriously sped away.

"What was all that about?" I asked.

Jeff explained. "He asked, 'Are you journalists?' 'No,' I said, 'geologists.' 'Good!' he said, 'We don't like journalists.' "

But the most interesting encounter of all was yet to come. By mid-afternoon we had reached a small road that extended south from the main section of Wadi al Jawf. Along this road were a series of younger sediments that again looked more weathered and less likely to preserve fossil bone than had been suggested by satellite images and geologic maps. "More Holocene—young stuff," Ian muttered in frustration. "Well, let's have a look anyway."

Enjoying the leg stretch, I wandered back down the road from the others, inspecting the same drab cuts a few hundred yards from our road stop. I looked northward to the flat, sizzling center of Wadi al Jawf, its surface tormented by whirling dervishes of dust. Then I noticed that one of these devils was getting closer. Soon I could see a small black speck at the bottom of the cloud. Within moments the speck became a car, wobbling like a mirage in rising waves of heat. Soon it was clear that the car was, predictably, a Toyota Land Cruiser, but one bristling with armament. I thought of that scene in David Lean's movie *Lawrence of Arabia*, where Lawrence witnesses a tribesman on horseback appear on the distant horizon and calmly shoot his Arab escort from some immeasurable distance for trespassing at his well. I stood transfixed as the vehicle came upon me, made a wild right-hand turn, and came to rest in a cloud of Holocene dust. The man in the driver's seat got out and walked purposefully toward me. He stopped within inches of my face and gave me a grimace and an interminable study.

"Where are you from?" he asked in perfect English, to my amazement.

"New York," I said.

"Pleased to meet you! I'm from Detroit." He laughed and shook my hand. "My name is Mana W. Ashaif." He then waved his hand in a respectful way toward a young, regal-looking man in the front seat of the Land Cruiser. "This is my sheikh, Mohammed Naji Abdul Aziz Shayif. We are all going to a party at my new house. We would like you and your friends to come."

I walked casually over to my comrades, still frozen in their tracks, and gave them the news of our invitation.

Ali responded with his usual advice. "Oh no, Meester Mike, these are bad men, they will—"

Before Ali could finish, Ian interrupted and issued a polite but firm command. "Ali, these fine gentlemen have offered their hospitality, and I don't think we are in a position to refuse them."

Ashaif had invited me to ride in their car, so I squeezed myself in the backseat between three warriors and four automatics. We sped out over the broad flats. The men laughed and whooped as if they were attacking a fortress full of infidels. More dust devils kicked up by racing vehicles appeared alongside: first two Toyotas, then a half dozen of them. We all raced toward a distant hill crowned not with a citadel of defenders but with the impressive Ashaif estate. I could not imagine a better Toyota commercial for an Arab audience. When we reached the parking area below the house, a few Toyotas were already parked and there was a flurry of activity. A woman drove up and dropped off her husband, waving her finger at him and warning him not to stay out too late with the guys.

The Ashaif house was another architectural wonder. Its clean limestone walls were studded with luminescent stained-glass reds, yellows and blues.

"Nice place," I offered.

"I had a very successful vegetable market in Detroit," Ashaif explained.

Thirty-five tribesmen dropped thirty-five automatic rifles and thirty-five pairs of Korean-made thong rubber sandals on the floor of the *mafraj*, the social room of the house. The sheikh asked Jeff and me to sit next to him in the corner of this stately room, next to some open windows. Here he did his official business. He was young—in his late twenties—but obviously given due respect and authority while his father was away in London. Two men came in and stood on opposite sides of the sheikh while he adjudicated their dispute and then dismissed them. Meanwhile *qat* was passed around in long troughs of sliced bamboo, and offered with more than the usual ritual. After receiving an Arabic explanation Jeff re-marked, "Expensive stuff! That tube is worth about a thousand dol-lars!"

It was both impolite and self-depriving to refuse an offer of a chew. Ian complained that even this high-grade stuff provided nary a buzz, but Jeff and I claimed to feel something. I looked out the windows to an ellipse of pearly sand, a series of black mountain peaks, and an azure sky, and felt I was drifting into a painting. Yet the trance seemed rather subtle. Jeff said to Ashaif and the sheikh that this seemed a safer combination with all those guns than whiskey. The sheikh laughed but replied that guns were all very safe: their use was governed by very strict rules that reflected the symmetry of Arabian law, and would be judged anywhere on a spectrum from contemptible, forbidden, reprehensible, neutral, rec-ommended, to obligatory. He then spoke of events and history that clearly reaffirmed the mastery and independence of the people of the Jawf. "This is the second-largest tribe of the Jawf. My ancestors took San'a twice, most recently in 1948." When he asked about our business, the sheikh was charmed to learn we were looking for the bones of dinosaurs and other ancient beasts, and not oil or gold. "I'm sorry we have too much oil and gold and not enough bones," he laughed.

As the party carried on, Ali remained greatly agitated. He mut-tered something to Ian about "very bad men" and begged us to make a hasty exit and drive the sixty miles to the capital city once sacked by the sheikh's ancestors. Realizing that his entreaty was having no effect whatsoever, he skulked out of the *mafraj* and stood on one of the grand porches of the house, chewing *qat* and talking with some of the outsiders awaiting a court session with the sheikh. We at last said our fond farewells, with expressed intentions to reunite. I was convinced that in the event of another invasion of San'a by this fierce but friendly Jawf tribe, the sheikh and his men would certainly be welcomed at AIYS. Ironically, we now had dif-ficulty extricating Ali from the scene: he had worked up a consid-erable wad of *qat* and was deep in animated conversation with his companions on the porch.

Subsequent events worked in Ali's favor. After some fruitless wandering in a travertine canyonland, we were harassed yet again at a military checkpoint where the guards searched for weapons, of all things. This foolishness wasted most of the late afternoon and we ascended an escarpment of Amran searching for a campsite. Instead we found another military checkpoint, whose soldiers informed us we were on the wrong road, one headed for San'a instead of our targeted field area around a large volcanic outcrop. To Ali's muted delight, many miles and meddlesome checkpoints later we reached San'a, just before midnight. At the AIYS we raided our secret liquor cabinet, swigged from our one precious bottle of Johnnie Walker whiskey, and reflected on the strange events of the past few days.

The first two days of July brought some fossil rewards. Dr. Hamed Nakhal took us to his quarry of fossiliferous shales twenty miles north of San'a. These rocks were part of the so-called transitional series of rock layers that bridge the gap between the Jurassic Amran and the Cretaceous Tawilah. It took only a few minutes to split open shales that contained some fine fish skeletons. We exuberantly started chopping into the quarry, hoping to find other evidence of Mesozoic life—frogs, dinosaurs, flying reptiles or pterosaurs, whose remains have been found in such fish quarries in other parts of the world. We even had hope for a long shot—bones of small Mesozoic mammals. But such discoveries seemed highly unlikely, given our luck thus far. The day ended with a fine Mesozoic fish fry but nothing else. As we prepared to leave the quarry a group of nagging kids broke into the Toyota and grabbed Ian's binoculars. With much difficulty we retrieved them and made a hasty exit.

It was now July 3. We had only ten days left in the stubborn geology of Yemen. Early on the morning of American Independence Day we got word that the U.S. Navy had inadvertently shot down an Iranian passenger jet, killing all 250 aboard. We paid a visit to the Ministry seeking some counsel about the situation. One of Ali Jabr's aides who met with us treated the whole affair in a surprisingly casual manner. "It is sad, but these things happen in war. Besides, you know, we are Iraqi allies in this war. Still, there are Iranian sympathizers, especially in San'a. I suggest you boys go to the country for a few days."

This meant only one option. We were still waiting for some communication from the oil geologists in the vast sand desert east of the town of Marib, our last unexplored area, so we decided to return to the unattractively sweltering Tihama. The second trip was as uncomfortable and unproductive as the first. Back in San'a on July 6, we headed west again on the 7th, this time finding some impressive fossil tree stumps in a bed of siltstones. But there was no bone anywhere. I remembered Vaughn's old adage that fossil plants and bones don't mix. It seemed a shame to have to travel halfway around the world to the exotic Arabia Felix just to prove his point. We continued up a wretched road full of craters and flood channels that crested at the summit of 12,000-foot Jebel Nabī Shu'ayb, the highest point on the Arabian Peninsula. From that stupendous overlook we could look west and see the haze of a dust storm on the horizon over the Sahara. The Yemen Highlands were an enchanting world for everything but vertebrate paleontology.

All this wandering left only the final trip to the desert near the ancient town of Marib. On July 8 we loaded up our road-ravaged Toyota and headed east, dropping down the awesome 3,000-foot escarpment of Amran limestones to the desert floor. After getting lost in the dunes, we at last found our way to the Hunt Oil Drilling Camp Number 5, where we were treated to a shockingly bizarre American meal of ham, corn on the cob, Coke, and French fries. We spent the next day wandering around the sand dunes, following the lead of oil geologists who claimed to have stumbled into small blowouts between the dunes covered with a pavement of small fossil bones. But these again proved to be very recent indeed; they were merely carnage sites of late Paleolithic or early Neolithic times. Only a few thousand years old, these little gullies were archaeological, not paleontological, attractions. We found some nice arrowheads, fishing points, and bones of domesticated sheep and goats. I picked up a pile of these objects but casually threw the stuff back on the ground before stepping into the departing vehicle. Our hosts from the Hunt team were crestfallen with our lack of enthusiasm. They had treated us so well, feeding us huge meals and help-ing us navigate around in the trackless dunes. They encouraged us to keep trying at other sites, but we had a good sense of the im-poverished lay of the land.

Marib itself was not so hospitable. This town was a faded rem-nant of the once-great city of the Queen of Sheba, who ruled the Arabian world and seduced the land of Israel and its King Solomon with her beauty and intelligence. A few eroded monuments of this past glory could be seen poking up through the sand, but archaeol-ogists working in the area complained of hostile treatment by the locals. We stopped at a gas station that lacked electricity for the pumps. A sullen teenager toting an AK-47 said we should wait an hour or so for the generator to kick on. To ensure our patience he refused to reinstall our gas cap. Inside the station were a group of equally insolent, xenophobic, and well-armed youths. Things be-came very tense, but at last the pump jockey produced the cap. We grabbed it from his clutches and swiftly departed.

Helping us in this confrontation was another Ali from the Min-istry, Ali Shaker, who seemed to have much more composure and valiance than his predecessor. He now guided us to the one gas sta-tion in Marib that seemed to be working, where we queued up in a line of rumbling cars. Armed drivers blasted their horns in macho threats of potential combat for a place at the pump. As I stepped out of the car, a Toyota truck blinked its headlights, and I saw three young men in the cab sneer at me. At last we got *malion*—a full tank—and eagerly left Marib in the yellow dust. Ali Shaker said, with more credibility than the previous Ali Mohammed in the Jawf, "This is a bad place with many bad people."

This was also essentially the end of the 1988 Yemen Paleontolog-ical Expedition. I knew we would have a few remaining days for prospecting in the areas surrounding San'a, but these hardly held much hope to reverse our misfortune. I was gloomy with this thought as we drove over the volcanic plain of Marib, past the dunes and the broad wadis sprinkled with rusty brown acacias, to-ward the huge escarpment of Amran, which blocked the sands from drifting west. At the top of the cliff the air cooled slightly but the sky seemed as heavy and dark as my mood. The mountains to the west were merely translucent shadows against the gray haze of the sky. Like their insubstantial appearance, these mountains seemed almost memories. We passed more enchanting stone villages that sprouted like rock crystals from the ridge crests. Ali talked about his Allah, his love of nature, and his sense of ethics. I wanted to know more. Renegade armies, bandits, and truculent teenagers notwithstanding, I had become seduced myself by the land of Sheba. I yearned for a paleontological reason to come back, to try once again to plunge into the great sandstone canyons and to wander the Wadi al Jawf. But nothing we had found over the past five weeks encouraged any expectation of return. I looked upon the dazzling Cretaceous cliffs northeast of San'a for what was probably the last time. At that moment I had to acknowledge a keen sense of clear and utter failure.

23

. . .

To the Foot of the Flaming Cliffs

In the months following our unsuccessful expedition to Yemen, Ian and I grew less daunted by the prospects of return. We felt that National Geographic might be willing to foot the bill for another attempt, and those fish we had found with Professor Nakhal were enticing. John Maisey, a colleague at the museum, identified these broadly as ray-finned fishes (leptolepids), definitely marine, and either Late Jurassic or Early Cretaceous in age. John ended his note on the fishes by remarking, with British understatement, "I would very much like to see more of this very interesting material."

Other news of interesting fossils soon came out of Yemen. Hamed wrote that a slab of volcanic ash containing some exquisite, complete frog skeletons had been pulled out of a stone quarry at El Rayaashyah, south of the town of Radā. He brought the material with him on a visit to the United States in the summer of 1989. The frogs, as far as we could tell, belonged to the genus

Miopelodytes, probably about Miocene in age, about 20 million years old. Perhaps there were some mammal skeletons buried in one of those piles of fossil-laden ash that had filled an ancient lake. Perhaps with enough time, hard knocks, and effort we might squeeze some interesting fossils from those Arabian stones. But by the time Ian, another National Geographic grant in hand, was ready to head out for Yemen again in 1991 with Peter Whybrow from the Natural History Museum in London, I declined and asked Jim Clark, a postdoctoral fellow at the American Museum and a talented young paleontologist, to take my place.

The reason was not a wave of sobering reflection about our past frustrations in the land of Arabia Felix. I was already much preoccupied by another spot in Asia—this one more gloriously productive and more than 8,000 miles away from Yemen. The unexpected opportunities for this venture had actually emerged in early 1989, less than a year after our return from the Arabian Peninsula and only few weeks after a more successful expedition to Termas del Flaco, high in the Chilean Andes. Like other such opportunities this one began with a communication and a visit. A delegation from Mongolia had been permitted to tour the United States—a sign of relaxation of the vise grip that the Soviet Union held on this remote, landlocked country. Dr. Sodnam, president of the Mongolian Academy of Sciences, requested a visit to the American Museum in New York, with the intention of viewing the fossil treasures retrieved from Mongolia's Gobi Desert in the 1920s by the museum's teams led by the legendary Roy Chapman Andrews.

In 1906 Andrews had come to the American Museum with only a bachelor of arts degree but with big ambitions. Desperate to land a job there, he promised to do anything, even mop floors, to gain entry. Hermon Bumpas, the museum's director, took him at his word, and Andrews was soon enthusiastically cleaning the floors of the taxidermy lab while he improved his technique as a taxidermist's apprentice. Soon he graduated to cleaning stinking whale carcasses stranded on the Atlantic coast, anything to fill a useful niche at one of the world's great institutions for research in paleontology, zoology, and anthropology. It was also for the American Museum that the great anthropologist Franz Boaz first explored and collected cultural artifacts in the far Northwest of North America and the isolated villages of Siberia. Through Boaz's efforts and his scholarly descriptions of Native American cultures, modern anthropology was born; it was further developed by one of his famous students, Margaret Mead, also at the American Museum. Then there were those dinosaurs, delivered to the museum in train cars from legendary localities like Como Bluff. Barnum Brown, William King Gregory, William Diller Matthew, Henry Fairfield Osborn—all great men of paleontology—had been or were museum employees. Andrews soon got the nod to do a little exploring himself. After many months in Japan, wined and dined in palatial brothels on Yokohama's Yoshiwara, or "Street of Joy," he traveled, like a twentieth-century Marco Polo, to China, where he savored the luxuries of Beijing and led a small expedition to collect birds, reptiles, and mammals in the mountains of Yunnan Province.

After those early wanderings, Andrews was hooked on the Far East and especially on the mysterious core of the Asian continent. In 1920 he persuaded the museum's president, Henry Fairfield Osborn, to support an audacious plan. A number of theoretical papers had suggested that Central Asia was the fountainhead of mammalian life, which subsequently radiated and dispersed all over the globe. According to several highly respected experts like William Diller Matthew, this mysterious core of the world's greatest landmass was the wellspring for rodents, various herbivorous mammals including horses and camels, primates, and, perhaps most provocatively, humans. Yet there was very little direct fossil evidence to support these theories. Andrews asserted that fossils had to be found in those wastelands to validate these important hypotheses.

This appealed to Osborn for both scientific and less distinguished motives. Osborn did not like the idea that humans might have evolved from African populations—he was a racist and a racial supremacist, an attitude evidenced by his presidency of the Society of Eugenics. Like his fellow eugenicists, Osborn advocated the suppression of reproductive vigor in the less fortunate sectors of society. The idea that humans ultimately evolved from the peoples of Central Asia was far more palatable to him than the alternative that modern humans all arose from populations in deep and, most important to him, dark, Africa.

Thus Osborn was predisposed to encourage, for partly despicable reasons, Andrews's rather harebrained scheme to traverse the void of the Gobi Desert for evidence of early man. The Gobi was not trackless; for centuries a few important camel caravan routes had connected the Mongols with China and the Silk Road. But beyond those few transects, there was nothing but sand dunes, pebble plains, barren mountains, and a few strangely resilient nomads—certainly no sign of fossils that might indicate that Andrews had the right hunch. Nonetheless, Osborn and a few other wealthy New Yorkers, including the imperious J. P. Morgan, bankrolled the risky venture.

The first expedition set out from Beijing in 1922, reached Kalgan, near the Great Wall, and pushed north into the unknown. This was an elaborate and expensive affair, which involved the use of automobiles for the first time in any scientific expedition. It was further planned that the three Dodge touring cars and two Fulton trucks would eventually be joined by some one hundred camels carrying food, fuel for the vehicles, packing materials, and, if things worked out right, fossils. Not far north of the border between China and Mongolia, the expedition started turning up bones of fossil mammals, fragments of assorted limb bones and the tooth of an ancient rhinoceros. The expedition's scientific leader, Walter Granger, proclaimed in jubilation, "Well, Roy, we've done it! The stuff is here." Nonetheless, it soon dawned on the team that spectacular finds—including any fossil evidence for early humans—would be more elusive.

The expedition endured major trials and tribulations—sandstorms, routes blocked by quick mud, bothersome bandits, and political intrigue. By September 1, 1922, they were still stumbling around, actually a bit lost, on a grassy plain just north of an imposing mountain range known as the Gurvan Saichan (Three Beauties). While Andrews stopped at an isolated ger—the Central Asian nomad's dome-shaped tent—for directions, the team photographer, John B. Shackelford, took a stroll toward what looked like the edge

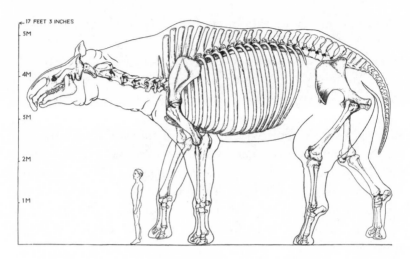

Indricotherium, formerly called *Baluchitherium*, a rhino relative and the largest land mammal, found in Mongolia by the Central Asiatic Expeditions under Andrews. (Their reconstruction from 1935 is slightly overestimated in scale.)

of the plain. This rim was in reality an impressive drop-off of red sandstones carved into walls, towers, and balancing rocks. Shackelford crouched down to examine the surfaces of the slopes and immediately started finding bone.

The team then set upon the rocks and came up with fragments of dinosaur skeletons, small bones, and even some bits of structures that looked like eggshells. Oddly, their time at this intriguing place was shockingly brief. In the Gobi, by September the weather already becomes erratic and an ill wind from the Siberian north heralds an excruciatingly long winter. After only a few hours of afternoon prospecting Andrews ordered the retreat of the caravan to China, but he resolved to make a beeline back to the red rocks in the following year.

That commitment to return to the isolated outcrop was a very good idea. The 1923 Central Asiatic Expedition turned out to be one of those glorious enterprises that vindicates all the failures and setbacks—all the bone-dry bouts in Yemen or El Rosario Baja—that are so much a part of paleontology. Soon after reaching the Gurvan Saichan the team settled in at the red cliffs, which Andrews called the Flaming Cliffs but were known to the nomads as Bayn Dzak because of the many (*bayn*) scraggly *dzak* trees rooted in the

flats at the bottom of the escarpment. During this season the Flaming Cliffs yielded scores of the beak-headed dinosaur *Protoceratops*
and the first associated nest of dinosaur eggs known to science. For
these discoveries the expeditions were acclaimed worldwide not
only by the scientific community but a public ravenous for news
about dinosaurs and dinosaur nests in exotic places.

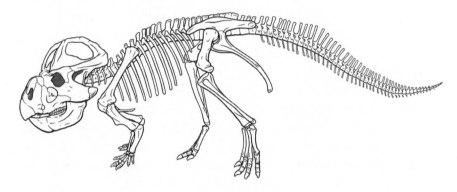

Protoceratops

One triumphant field season followed another. The American
Museum team uncovered not only more *Protoceratops* and their supposed eggs, but evidence of the ferocious little predator *Velociraptor*
and the very odd, birdlike predator *Oviraptor*. In what many regarded as the most scientifically significant finding, the paleontologists scooped up the skulls of small mammals near where dinosaur
skeletons had been found. Some of these skulls were no bigger
than an almond, but they were extremely important. Up to that
time very little was known about mammals of the Cretaceous that
lived alongside dinosaurs, and most of the evidence hailed from the
American West, from Wyoming and Montana, where these tiny
creatures left their remains as enigmatic fragments of teeth and
jaws—as we call them, "spare parts." The exquisitely preserved
skulls of Cretaceous mammals from the Flaming Cliffs—animals
that received tongue-twisting names like *Zalambdalestes*, *Dja-
dochtatherium*, and *Deltatheridium*—offered insights into the early
species that foreshadowed the great evolutionary radiation of the

more modern mammal groups recorded in the fossil record just after the Cretaceous extinction.

The Central Asiatic Expeditions thus had a series of major accomplishments wrenched from laborious and even dangerous exploration. Many of the discoveries were wholly unexpected. After all, Andrews and company had made the serendipitous discovery of one of the world's greatest dinosaur and ancient mammal sites, the Flaming Cliffs, because they were lost and had had to stop to retrace their route. And in the end, Andrews did not deliver on the promise that launched his expeditions in the first place. For all their tireless searching, the Central Asiatic Expeditions failed to find a scrap of human remains more than a few thousand years old. But in the course of this failure, the team enjoyed marvelous and paradoxical successes, a pattern common in paleontology. Those dinosaurs and ancient mammals from the Flaming Cliffs and other localities have had an enduring effect on our understanding of the evolution of life during the Cretaceous and Early Cenozoic.

This succession of paleontological victories had a downside. Through the discoveries themselves and a relentless campaign of public promotion, Andrews attracted a lot of attention—not all of it good. It was true, as he claimed, that such publicity was necessary to keep his elaborate enterprise financially solvent. But it was also

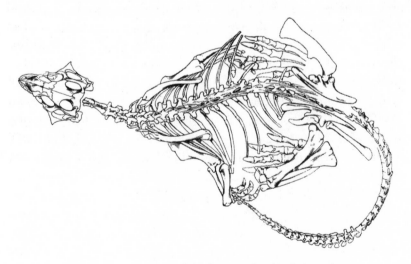

Psittacosaurus, a primitive Gobi dinosaur related to *Protoceratops*

evident that the fanfare was at times excessive. The attention was focused on the romance and adventures of Andrews the explorer rather than on the scientific achievements of the expeditions, and the publicity stunts were too contrived. One scheme involved an auction of a putative *Protoceratops* egg, a move that Andrews recalled later with bitter regret. The auction suggested to the Mongolian authorities that this material had very high monetary value. This suspicion does not seem far-fetched, given the premium today placed on some fossils—a big *Tyrannosaurus* named Sue (which auctioned in 1997 for several million dollars), pterosaurs, fishes, and ammonites for sculpture or house ornaments and even dinosaur eggs. In the 1920s, however, such fossils had very little market value, and Andrews was now plagued with a problem he had created. The governments of Mongolia and China became more scrutinizing and controlling about access and exportation. To make matters worse, the ideological warfare that pitched Chinese against Russians and divided the Mongolians was in full swing. Andrews was thwarted at every turn in his negotiations and his series of outstanding scientific expeditions did not outlive the decade.

Andrews enthusiastically spoke of returning to Mongolia once the situation improved, but events of the subsequent decade precluded this. The victorious Mongolian-Soviet alliance severed all diplomatic connections with China. Mongolia's capital, Urga, became Ulan Bator ("Red Hero"), and the country itself fell into line as part of Stalin's oppressive empire, aided and abetted by a bloodthirsty puppet dictator named Choibalsan. One of the "highlights" of Mongolia's history in the 1930s was Choibalsan's purge of Buddhists, which resulted in nearly 30,000 deaths and the destruction of all but four of Mongolia's 700 Buddhist temples. Meanwhile the fossil badlands of the Gobi lay untouched.

It was not until 1946 that a well-planned Soviet expedition under the direction of the famed paleontologist Ivan Efremov once again penetrated the desert. Efremov used heavy Soviet military vehicles and targeted places that had been inaccessible to the 1920s expeditions with their spindly-wheeled Dodge touring cars. The Soviets reached the white-hot core of the Gobi, a place called the Nemegt Valley, a scorched basin between two mountain ranges located less than eighty miles north of the tense Chinese border. There, the team uncovered new marvels of the age of dinosaurs in Central Asia. The findings included a magnificent *Tarbosaurus*, a dead ringer for a close relative of *Tyrannosaurus*, as well as large aggregates of duck-billed dinosaurs like *Saurolophus*, found in evocatively named quarries like the "Tomb of the Dragons." Soviet activities continued intermittently through the decades, sometimes with the collaboration of Mongolian scientists trained in Moscow.

Joint Polish–Mongolian expeditions carried out in the late 1960s and early 1970s were successors to this melding. Under the leadership of a young Polish scientist, Zofia Kielan-Jaworowska, these expeditions triumphed on many fronts. The scientists found remarkable dome-shaped dinosaurs like *Stegocephale* and tiny mammals like *Asioryctes* in a series of red sandstones of Nemegt Valley that looked so inauspicious that the Soviet had named it the Barren Zone. Kielan-Jaworowska and her team also crawled on their hands and knees at the bottom of the Flaming Cliffs and came up with some precious new mammals like *Kennalestes*. Less than sixty miles from the cliffs they scoured a lone butte, Tugrigeen Shireh. This escarpment sandstone was not red but blinding white in color, although its sediments preserved *Protoceratops*, small mammals, and other forms identical to the fossils at the Flaming Cliffs. Here, the paleontologists found something remarkable—a *Protoceratops* and a *Velociraptor* intertwined in obvious combat—one of the most spectacular and evocative fossils ever found.

This was all good work, but by 1989 the Gobi as fossiliferous territory was still worth a lot more examination. Since Andrews's era the Gobi had been crisscrossed by expeditions sanctioned by the governments of the Soviet empire, but these explorations hardly compared in intensity with that of other great fossil territories such as the basins, ranges, and canyonlands of western North America and Argentina. And scientists from the West barely had access to important specimens in the Soviet collections, let alone the privilege to roam the Gobi for their own discoveries. As relations with China improved, Canadians conducted a series of productive collaborations in the part of the Gobi Desert that extends to Inner Mongolia, a region under the control of the People's Republic of

China. But this southern terrain represented less than a third of the fossiliferous land area of the Gobi.

With this history in mind it is easy to see why a visit at the American Museum from an undersupported scientific academy in an isolated, landlocked nation might have more than passing interest to us. At the museum, Dr. Sodnam and his entourage marveled at the amazing series of skeletons of *Protoceratops*, the famous beaked-faced, frill-headed dinosaur that the American Museum team had unearthed in 1923 along with the nests of dinosaur eggs. The 1920s expeditions indeed had recovered more than one hundred skulls of "protos," and these were an integral part of the world's largest dinosaur collection. Inspired by the tour, Dr. Sodnam extended a remarkable invitation, one that I never expected to hear in my paleontological lifetime.

"Why don't you come back?" he asked, thereby casually erasing the sixty-year moratorium on any work in Mongolia by any Western paleontological team.

"Is this possible? Why . . . yes, sure," I replied, a bit in shock.

"Of course, there are the necessary preliminaries, but I believe an expedition could be arranged," he continued in a more realistic tone. "Can you come this summer?"

That moment had an eerie resonance I couldn't quite define. Later I realized what it was: it was like being reborn. It felt a bit like that first invitation to real field paleontology that Peter Vaughn had extended to me nearly twenty years before: "Would you like to go in the field?"

"Why . . . yes!" I exclaimed.

After Sodnam's departure, I confess I forgot about the whole matter. It seemed unbelievable. How would the Soviet leadership, who jealously guarded their satellite countries, allow such an invasion from the West? Thus it came as a great surprise when I heard from the Mongolian Academy several months later. I was in a Buenos Aires hotel on a cold August day when I received a phone call from New York. My assistant, Barbara Werscheck, gave me the stunning news: "The Mongolians have sent a message. They ask when are you coming, they are waiting and ready to take you to the field." We sent a message back indicating our willingness to join them at the drop of a hat. But that was the end of communications for several months.

Then in January 1990, with the Soviet Union in near collapse, Mongolia declared its political independence. A small delegation came to the museum that same month and met with Malcolm McKenna and Mark Norell while I was out of town. The Mongolians followed up with an official invitation. I responded with a query for an estimate of our costs for a small reconnaissance— about three or four people. But communications with the Mongolians did not seem much better than they had been in the time of Andrews. With only sporadic news from Ulan Bator, and without any guarantee of being able to get train reservations from China to Mongolia, Mark, Malcolm, and I, along with Aldona Jonaitis, the museum's vice president for exhibition and public programs, left for Beijing in early June, at just about the time of the first-year anniversary of the tragedy in Tiananmen Square. Our friends at Beijing's Institute of Vertebrate Paleontology and Paleoanthropology managed to secure us seats on the Trans-Siberian Railway and we were off into what Andrews often referred to as "the great unknown."

That first train ride itself was as memorable as any adventurous old-fashioned caravan into the Gobi. We were comfortably lodged in a pleasant first-class compartment trimmed with mahogany and brass. Our carriage was shared with a group of rowdy American fishermen who had nabbed a special deal to gain entry and guidance for sportfishing in the brilliant blue lakes and sparkling rivers of northern Mongolia. These guys knew how to party. One of their several reserved compartments was entirely filled with cases of beer. I looked out the grungy window and for the first time saw China—the bronze-colored brick-and-adobe buildings, the hills pockmarked with ancient tombs, the power lines, and the steam locomotives. There were throngs of people everywhere, carrying enormous loads of firewood or water buckets, serenely drifting by on bicycles, towing wheeled carts full of rebar, pipe tubing, or chicken wire, or driving odd three-wheeled contraptions that looked like oversized lawnmowers, which we nicknamed "Toros." As we crossed the steep, forest-clad mountains north of Beijing we soon confronted the Great Wall itself, snaking over summits and ridges and plunging into steep defiles. I thought of Andrews and his jumping-off point at Kalgan, near the Wall. How familiar the whole scene even now might look to him.

As we moved north of the Great Wall an ineffable atmosphere, perhaps the effect of afternoon light bent in a prism of dust, suggested that the Gobi was near. By this time some of the fishermen were quite drunk, regaling us with tales of their exploits in China—including their fun at the BK Gun Range, a sort of amusement park for marksmen run by the People's Liberation Army. George, from Atlanta, one of the most loquacious and inebriated of the lot, exclaimed, "It's great—you can shoot a round of anything—antiaircraft, antitank. You can shoot a bazooka and blow up a wall. They'll even throw a few chickens out there for the sport of it."

Drifting away from this animated cocktail hour, I moved to an open window at the back of the car where I could stand in solitude. I was suddenly struck by the smell of the desert, this strange desert more than 13,000 miles from my home but a desert with a smell as familiar to me as that in New Mexico or Baja California.

Our first entry into Mongolia at the border town of Erlian was an elaborate ritual that seemed appropriate for the importance of the event. Late in the evening, to the sound of shrill Chinese music and a scratchy, indecipherable series of announcements, all passengers were offloaded and ushered into the station while the train was pulled into a vast, hangarlike structure. Here, an energetic crew of mainly Chinese women manipulated heavy machinery to lift the carriages from the tracks and replace the wheel sets with the gauge suitable for Soviet-built rails. Meanwhile we congregated in a large hall that emitted an overpowering stench of urine. We queued for a passport check with Chinese and Mongolian traders, embassy staff, Russian students, a Chinese chess team, bedraggled backpackers from East Germany, missionaries, and our listing fishermen friends. This involved an interrogation in Chinese that we could neither understand nor respond to. Fortunately, there were some instructions in English—written in block letters, many misspelled, like a second-grade homework paper—hung like a giant mural on the wall. The instructions were detailed and intimidating. In an expression of sympathy the sign ended with the words:

DON'T WORRY

FILL FORMS

CHECK PASSPORTS

THEN, BAR ON LEFT

The Erlian pageantry took more than two hours, and it was well after midnight when we reboarded our train with its new wheel sets. I awoke several hours later from a sleep intensified by the rocking rhythm of the train. The tracks wrapped around low hills carpeted with the silver grass of the Asian steppe. Through the window I could see the snub-nosed diesel locomotive rounding a curve so extended that it looked as if the train were circling back toward me. I could hear Mongolian children in a compartment near ours laughing and yelling *"Temi!"* A two-humped camel on a hill, my first sighting of such a beast, offered a translation for my first Mongolian word.

In the late morning we reached Ulan Bator and were received by a delegation from the Mongolia State Museum, where the di-

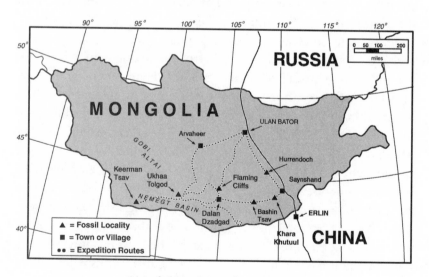

Major fossil localities in the Gobi Desert

nosaurs—including the famous fighting pair—were displayed. This turned out to be the wrong welcoming party—they were interested only in our giving them exhibits or helping them build one—and our intended scientific collaborators had missed us. After the proper meeting with the Mongolian Academy scientists, however, we holed up for days in the pseudo-luxury of the Ulan Bator Hotel. Here the rooms were magnificently airy, with soft, elegant couches and mahogany tables and chests, but the hallways reeked of sheep fat from the kitchen, the elevator was dysfunctional, and the toilets flushed as forcefully upward as downward. After several days, our colleague, the senior paleontologist Demberylin Dashzeveg, appeared and simply announced, "We go!" The next afternoon we were on the road to the Flaming Cliffs in two Soviet-built, heavy-duty, four-wheel-drive military trucks, not unlike the metal dinosaurs that Efremov first piloted to the Gobi in the 1940s.

By June 24 we were stopping a half mile or so from the cliffs to enjoy them from afar, but with a proximity that a Western paleontological team had not enjoyed in sixty-five years. The cliffs looked like all those illustrations from children's books I had pored over as a kid in Los Angeles. We prospected for a couple of days, turning up a few protos, ankylosaurs, and even some fragmentary bits of mammal bone. Then we struck south, out on a wild, rather disoriented

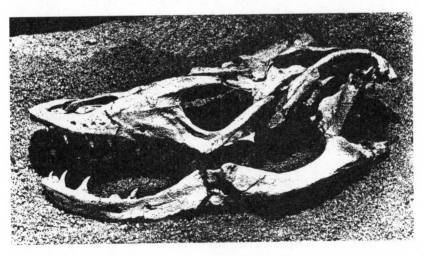

A photograph of the skull of *Estesia*, a varanoid lizard

trip over the lofty Gurvan Saichan, across dune fields, and into the center of the Gobi, the Nemegt Valley.

Even this brief jaunt was enough to convince us that we were going to have a time of it. On the north side of the Nemegt Valley, near an intricately carved natural amphitheater of red sandstone that the Poles had nicknamed El Dorado, we found some very good fossils. Most brilliant among these was an eight-inch skull and a partial skeleton of a large lizard. Mark later identified this form as a varanoid lizard, a curious animal related to the Komodo dragon, the largest lizard alive today. The Gobi lizard, however, had a series of canals running vertically through the teeth; these were similar in design to the tooth canals in living Gila monsters, and it is likely that they also carried venom from poison glands. The animal, clearly new to science, was named *Estesia*, after Richard Estes, a friend, colleague, and world authority on fossil lizards who had recently passed away.

The discovery of this venomous Komodo-like dragon was the event that secured our commitment and the requisite support for a great Gobi adventure in the years to come. We returned to Ulan Bator, signed an official agreement with the Mongolian Academy of Sciences, and started immediately planning for the 1991 Mongolian Academy–American Museum Paleontological Expedition. In the years since then I have often looked back on that first field season in Central Asia. It seemed so casually planned, so small, so spare in equipment, so inexpensive, so brief, and so decisively successful. That fine morning in June when, near El Dorado, we stumbled on a weathered ankylosaur skeleton, some limbs of a small theropod dinosaur, a nest of eggs, and of course the skull of *Estesia*, was a particularly victorious moment. But what I remember most was seeing the Flaming Cliffs for the first time. They were a somewhat drab orange-brown on that overcast day, not burnished and incandescent red, as Andrews often described them. I wasn't quaking with joy or kissing the sacred ground. It felt, instead, eerily natural to be at the foot of those cliffs. They were familiar; they felt like home, as if I had merely walked around the block of my parents' house in Los Angeles and into the land of the dragons.

24

. . .

The Cretaceous Cornucopia

I looked around me and felt alone. To the south the terrain gently sloped toward the middle of the Nemegt Valley, a white pediment that rose on its far side toward a mountain range swirled in chocolate and maroon. To the north, caught in the long arc of the Gobi sun, were the rippled muscles of another mountain range, the Gilvents. Beyond the Gilvents was another valley with no name at all, this one broader and hotter and more trackless than the Nemegt. And farther still, north of the great Gurvan Saichan range, were the Flaming Cliffs, the world's most famous dinosaur site, about 225 miles from where I stood.

As I scanned the foreground I could see a series of scenic but unspectacular red-brown cliffs. I could also see I was not alone. A quarter mile away, high on a ridge near two promontories we called the Camel Humps, were Pete Makovicky and Diego Pol. I saw the swing of an arm and the glint of a sharp point of a pick-mattock as they chopped away at the sandstone around another *Oviraptor* skeleton. It was not a surprising discovery. Over the years we had found more than a dozen *Oviraptor*, including a couple of nesters—oviraptorids squatted on a pile of eggs—at that very same level we called Death Row. Closer to me, I could see Amy David-son intently scooping sand from a near-vertical face, where the fragile bones of a delicate, birdlike theropod dinosaur were buried. Jim Clark and his wife, the dinosaur paleontologist Cathy Forster, were in the broad, bone-dry wash just south of Ankylosaur Flats. Mark Norell was out of view at Xanadu, the fabulous site with nu-merous dinosaur nest sites and even eggs with the first embryos ever discovered in the Gobi Desert. Dashzeveg, our Mongolian col-league, was with his son again at Zofia's Hill. Not far from them, at First Strike, was the young paleontologist Bolortsetseg, doubt-less focusing her sharp eyes on the pebble-strewn terrain for small mammal skulls.

We were all there, at this place called Ukhaa Tolgod—dispersed, isolated, not talking to each other for hours on end, but closely al-lied by shared failures, successes, and intensity of interest. This had been our little world for the last seven years, our own place, the place we had discovered. It wasn't much to look at, nothing like the Flaming Cliffs, with its 400-foot escarpment and its rock towers, certainly not like the great red canyons of Khulsan, Ikh Khongil, Nemegt, and Altan Ula, which opened into the valley some fifty miles west of us. But inside the puny cliff faces, small flats, and iso-lated mounds of rock at Ukhaa Tolgod were fossils far more nu-merous and arresting than elsewhere in the Gobi.

I moved slowly downslope toward a set of low ridges in the flats. The surface had a few footprints, including the oversize prints of big Jim Clark. This was a broad apron of red sand below a slope we called Delta Force, where several years before I had found a little cluster of three precious skulls and skeletons of the important mammal *Deltatheridium*. I could also see other glory holes—one where Mark and Amy pulled out a small *Oviraptor* that year of dis-covery in 1993, another where a young, inexperienced Mongolian student made a mess of a weathered *Oviraptor*, leaving behind a pile of rubble. Here was the mound that was once ornamented with a fine *Protoceratops* skull. Here was the ridge where Jim picked up several small mammal skulls in a single afternoon. And here, on a little island of sandstone poking through the grass, we had once encountered a fine specimen of the big lizard *Estesia* as well as another lizard with its delicate little skull and its coiled trunk and tail not far away.

As the hours passed, Jim slowly drifted toward me. Not until he was within a few yards—less than the length of an average-sized ankylosaur—did he speak out.

"Hi, Mike."

"Hi, Jim."

"Find anything?"

"A few things—a multi skull, a theropod fragment. How about you?"

"A nice lizard and a couple of multis, and part of a mononykhid."

There was a long silence. Then I heard a muffled voice as Jim kept moving south, his back turned away from me.

"Bye, Mike."

"Bye, Jim."

I took a swig of tepid water and watched a cloud floating in from China cover the sun. It was the last paleontological season of the millennium. Sometimes, to pass the long hours while prospecting the slopes, I would try to recall details of each summer in Mongolia since 1990. Where were we a year ago today? What did we have for dinner that day? Who got sick that day? Who went home that day? Who visited us that day? I couldn't even remember the precise date in July 1993 when our lives changed with the discovery of this magnificent fossil locality. It was impossible to get the flicker of images, people, and events to stay put anywhere for long. Annoying songs—like Wayne Newton singing "Danke Schön"— would plague me; I tried to banish them with other songs or other memories. I wrote novels, composed movie scripts, poetry, and all the while kept my eyes to the ground, with the obsessive attention to rock, soil surface, pebble, and bone chip I had learned from Peter Vaughn three decades before.

There was much talk this year of celebrating the ten-year anniversary of the American Museum–Mongolian Academy of Sciences Paleontological Expedition. But in the end we could not muster enough energy and enthusiasm to contrive some great ceremony for the occasion. There were several suggestions. We envisioned a party at the Flaming Cliffs, complete with fireworks and champagne cooled in emergency ice packs, or cold drinks and fresh salmon flown in by plane to the nearby airstrip at the Gobi tourist camp. The plane would be full of museum trustees, ambassadors, and movie stars. One of the Mongolians exclaimed, "It will be a great party, like a victory celebration for Temujin!"

In 1226, the Year of the Dog, Temujin, alias Genghis Khan, probably saw the Flaming Cliffs. On his way south with a small entourage, the most expansive landlord of all time lingered near the Gurvan Saichan to hunt the elegant wild ass, the onager, which roamed the Gobi plains in large herds. On one of these forays, he collided with an ass and took a bad tumble off his horse. The Khan was a formidable warrior but he was now quite old—nearly sixty-six—and the injuries he sustained kept him in a *ger* near the Gurvan Saichan for several weeks. In the late spring he resumed his trek to northern China, where he punished and slaughtered some of the rebellious populations under his reign. This was his last act of imperialism. Never quite recovered from his injuries of the previous year, Genghis Khan died in 1227, about to make his way back across the Gobi toward his capital in Mongolia.

One wonders, of course, if the Khan and other parties roaming the Gobi encountered skeletons of dinosaurs and pondered what creatures leave behind such fearful and bizarre structures. The *Protoceratops* skeletons at Flaming Cliffs are indeed hard to miss. The beaks and shields of these animals can be seen from a hundred yards away protruding from a cliff ledge or a small mound of sandstone. Some have even proposed that *Protoceratops* skeletons were the inspiration for griffins—those half eagle–half lions whose images guarded caves of gold. Evidence for such a connection is sketchy at best, but it is interesting that the legend of the griffins emerged from Central Asia, where, at least in the Gobi, both gold and *Protoceratops* are notably abundant.

Some spots in the Gobi are surprisingly thick with skeletons.

Moreover, the preservation of these remains in soft, fine sands exceeds that of virtually anything found in western North America. The fossil sites uncovered by the original Central Asiatic Expeditions under Andrews and the subsequent expeditions led by the Russians, the Poles, and the Mongolians themselves are the paleontological equivalent of the great temples along the Nile. These Karnaks and Luxors have produced some of the world's most extraordinary concentrations of vertebrate fossils. In our hotel room in Ulan Bator in June 1991, we made a list of the sites that were potential destinations—the Flaming Cliffs and Tugrigeen nearby, Oshi for older dinosaurs and mammals, the orange sand ramparts in the Nemegt Valley at El Dorado and other sites within Ikh Khongil, the candy-striped cliffs of the Nemegt, the dinosaur beds of Altan Uula where the Russians excavated the Dragon's Tomb full of massive duck-billed dinosaurs, and to the west, the filigreed red cliffs of Khermin Tsav, which shimmered in the brutal heat of the desert. Farther west and north were the endless white badlands of Bugin Tsav.

These were the well-known places, most of them big badlands, with lots of exposed terrain. We knew that it was unlikely we would be the first to find such large-scale landscapes on our own. But some places, like the confined and unspectacular localities that

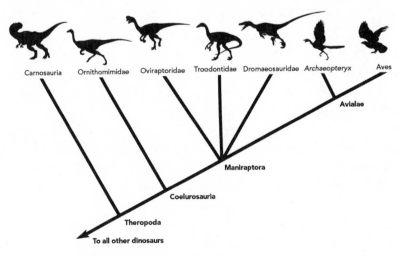

Relationships among the major lineages of theropod dinosaurs

flanked the Flaming Cliffs, were also productive. We reasoned that some pockets rich with bone would have been easy to miss in a virtually unmapped desert of half a million square miles—about the size of five Wyomings. With this opportunity in mind, we struck out in a series of far-ranging auto tours of the Gobi, scanning the poor available maps, satellite images, air photos, and the horizon for small islands of red rocks that might be replete with fossil bone.

But this wide-range reconnoitering did not pay off for some time. Although we found a variety of stunning specimens of *Velociraptor*, *Protoceratops*, dinosaur eggs, the flightless bird *Mononykus*, and the hopping mammal *Zalambdalestes* in the 1991 and 1992 seasons, we retrieved them all at known localities. We had not yet found the right new spot at the right time—the first time. There was after all nothing like the first time, when you walk into a gully strewn with skeletons and know it hasn't been walked before. Where were those localities? Had we entirely missed a mountain range and another valley beyond it?

On July 16, 1993, our third season in the Gobi, we had our answer. The choicest spot for Cretaceous fossils in the Gobi, if not the entire world—the Cretaceous cornucopia—was right under our noses. We had driven by this place three seasons in a row. Looking north from the center of the valley, all we could see were some bland, red-brown pimples of sandstone poking up against the black volcanic ribs of the Gilvent Range. The traditional name for this place said it all: according to Dashzeveg that cluster of small outcrops was simply referred by the locals as Ukhaa Tolgod, "Brown Hills." But in 1993 Dashzeveg and the rest of us decided to take a brief diversionary trip up the trail some ten miles to those isolated buttes. There, on the morning of July 16, we saw a field of red sediment covered with splotches of white. Each splotch was a dinosaur skeleton. By the end of the day we had found enough dinosaurs, mammals, egg sites, and lizard skeletons to rival the richest of all the Gobi sites, like the Flaming Cliffs, that had been scoured for more than seventy years. By the end of the second season we had accumulated specimens of 39 theropods, 260 lizards, and an unbelievable 187 mammals—more than in all the Gobi upper Cretaceous

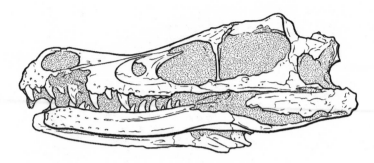

The skull of *Velociraptor*

localities put together. By the year 2000 we had lost interest in pre-cisely tallying these superlatives. During seven years of work at a site hardly larger than a quarter of Central Park, we had either ex-tracted or field-recorded at least 200 dinosaur skeletons and had collected nearly 1,000 mammals and 1,400 lizards, a haul that is enormous even compared with respectable results from many of the world's best dinosaur sites. But this numbers game is hardly in-dicative of the true significance of Ukhaa Tolgod. Many of those 2,700 fossils are so exquisitely preserved, so complete, and so fine in detail, that they have offered dazzling new insights into life in the Cretaceous Gobi 80 million years ago.

I have elsewhere related the trials and adventures of the Gobi expeditions and the triumph at Ukhaa Tolgod. Now, at the begin-ning of the twenty-first century, these experiences and findings have been greatly enriched by six additional years in the Gobi. Some of the new developments embellish the exciting dinosaur finds at the site made at the very beginning. That first glorious day at Ukhaa Tolgod in 1993 was rendered all the more dramatic by the discovery of a very unusual specimen. About midday, Mark came running back to the truck where I was counting specimens, ex-claiming something about how he had found the best fossil ever.

It was indeed very special. Mark took me to a flat where he pointed to a nest full of *Protoceratops* eggs. One of these was broken, and one could see that its contents—some small bones—adhered to the inside of the shell. "An embryo!" I shouted. Even more unusual were two tiny skulls of sharp-toothed creatures in the nest. These

An oviraptorid embryo reconstructed from the skeleton

were very young versions of meat-eating theropods, either drom-aeosaurids, forms that include the predaceous *Velociraptor*, or the large-brained troodontids, with bits of eggshell stuck to their snouts. We reasoned that these little skulls were probably associated with the creatures in the shell. This was big news. For seventy years, these oblong eggs, with shells sculpted into very distinctive small bumps and ridges, were thought to belong to *Protoceratops*. Now it seemed to us that these numerous nest sites actually belonged to a species of dromaesaur, a *Velociraptor*-like meat-eating theropod. But this field diagnosis was not quite accurate. Several months later, after preparation in New York under Amy's skilled needle, the true identity of the jumble of tiny bones in the egg was revealed. The embryo in the half shell was a baby *Oviraptor*, the animal originally indicted for stealing these very same eggs!

What, then, were the dromaeosaurids doing in that nest? The juxtaposition of dromaeosaurid and oviraptorid in the same nest seemed too unusual to be explained by chance. We conceived two possible scenarios: either the little dromaeosaurids raided the nest, or they were fed to the oviraptorid babies by mamma (or papa) *Oviraptor*. We favored the latter speculation. Whatever parental care was performed by the adult, it likely extended to feeding its voracious hatchlings.

There was even more riveting evidence at Ukhaa Tolgod of oviraptorid parental responsibility. On the second day at that site, Mark found a nice adult *Oviraptor* on a slope, its limbs splayed out into long finger bones and claws. Since we had found several such skeletons by then, it was not a sensational discovery. We did, however, select this oviraptorid as one worthy of excavation. A couple of days later we were stunned to see the purple sheen of eggshell exposed below the animal's rib cage and pelvis. The specimen was exposed in all its glory back at the museum, where it revealed itself to be the first fossil ever found of a nesting dinosaur. The specimen was evocative, with its low crouch, its forearms wrapped around an ordered pile of eggs in a protective embrace, and its hind limbs strongly flexed. This looked persuasively parental, as if the parent were incubating the eggs in much the same posture that we see in a living ostrich or a chicken.

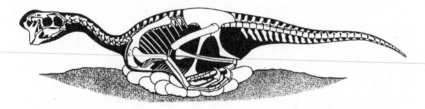

A nesting oviraptorid

The nesting oviraptorid, which we nicknamed Big Mamma, offered an insight into parental behavior that well matched our notion that these and other theropods were simply Cretaceous relatives of living birds. But was this simply a bizarre juxtaposition of an animal and a nest of eggs? There are several reasons for rejecting this suspicion. In 1923, the Central Asiatic Expeditions found at Flaming Cliffs an *Oviraptor* splayed out over a nest of supposed *Protoceratops* eggs. The name *Oviraptor*, or "egg thief," seemed appropriate for an animal caught in the act of raiding a nest. Now that the true producers of these supposed proto-eggs had been identified, the Flaming Cliffs specimen appeared to be another example, albeit a much less well preserved one, of nesting behavior in *Oviraptor*. In the early 1990s, the Sino-Canadian team found a very fragmentary fossil that also looked like a nesting oviraptorid. This specimen was retrieved from Bayan Mandahu, a large outcrop of red sandstones located about fifty miles south of the Chinese border that closely resembles the sequence exposed at Flaming Cliffs and Ukhaa Tolgod. Finally, we turned up a couple of more nesters at Ukhaa Tolgod among the scores of oviraptorid skeletons preserved there.

The large nesting oviraptorids of Ukhaa Tolgod are probably the best-known dinosaurs from the site. But others in this cast of Cretaceous characters are worthy of our attention. A more gracile form of oviraptorid is also well represented at the locality. In 1995 we found two of these complete skeletons lying side by side in deadly repose. The specimens—variously nicknamed for tragic couples, Romeo and Juliet or Sid (Vicious) and Nancy—were encased in plaster blocks we called "grand pianos" because they weighed about 800 pounds apiece.

We also found a variety of *Velociraptor*-like and troodontidlike forms at the site. One I uncovered in 1993 was a small troodontid represented by a slender skull with long jaws and a series of minute, sharp teeth. In 2000, this elegant specimen was described by Norell and his colleagues as a new dinosaur, *Byranjaffia*. These troodontids have a lot in common with birds, including their claw design, the presence of a crescent-shaped wrist bone (appropriately called the lunate), and an intricate labyrinth in the ear region for hearing and balance. At the sublocality Xanadu, we also found a nest of slender, pointed eggs, distinctly smaller than the oviraptorid eggs, that have embryonic remains of a troodontidlike form, possibly *Byranjaffia*.

The biggest dinosaurs at Ukhaa are the ankylosaurs, armored, tanklike beasts with spiked tails. Some of these forms are massive, attaining lengths of fourteen or fifteen feet. They also seemed to have congregated before dying. In one area we called Ankylosaur Flats, we had to step gingerly between more than a half-dozen ankylosaur skeletons to avoid breaking the bones. Such concentrations have been found elsewhere. At Bayan Mandahu, the Sino-Canadian team discovered a death assemblage of several ankylosaurs, both adults and juveniles, all oriented in the same direction.

The ankylosaurs were the bulky plant eaters of these Gobi sites. In North America, where dinosaurs of this size are often encased in cementlike rock, chiseling out one of these monsters could mean a whole summer's work. In Gobi sites like Ukhaa Tolgod, where the red sand surrounding the specimen is often very soft and friable, the excavation of ankylosaurs is a comparative lark. During the 1996 season, we conscripted the U.S. ambassador and a vigorous party of six volunteers to help us uncover a headless ankylosaur skeleton. The quarry team was as happy as kids in a sandbox. Within a couple of days they had trenched around the massive skeleton, excited and proud with their labor in the sun.

When extracting a big ankylosaur, a proto, or an oviraptorid nester, we have often chunked out whole mammal or lizard skulls. In addition, these specimens sprinkle the surface of various outcrops in extraordinary numbers. These range in size from the eight-inch skull of the lizard *Estesia* to walnut-sized skulls of a lizard called *Carusia* and some squirrel-sized multituberculate mammals. The smallest skulls—those belonging to either lizards or small insectivorous mammals—are tiny indeed, some not much bigger than a sunflower seed. To those accustomed to chiseling out big dinosaurs, these tiny fossils are easily missed, proverbial needles in a haystack. But the patient pursuit of such minuscule remains can yield surprisingly good samples. All it takes is the right glint of the sun's reflection off a flick of enamel. "Got one!" I shout. Maybe this is what it's like to be an entomologist swooping a net over a tiny leafhopper.

At Ukhaa Tolgod these small mammals and lizards are easier to find because there are so many of them. In fact, the density and fine preservation of the fossils there are every bit as astonishing as the abundance of sleek troodontids, nesting oviraptorids, and dinosaur embryos. Nowhere else in the world are lizards and mammals so well represented in the age of the dinosaurs. With this outstanding sample comes important new disclosures. Several species of mammals are preserved in fossils with such fine details of the pelvic structure that they even tell us something about their reproductive modes. A new form called *Ukhaatherium* and the long-legged, shrewlike mammal *Zalambdalestes* are represented by fossils with important details not found previously in Cretaceous mammals from either the Gobi or North America. For instance, both these animals have tiny struts of bone jutting out from the front of the pelvis. These struts, known as epipubic bones or sometimes marsupial bones, are primitive elements found in some ancient Mesozoic mammals as well as in living monotremes (the duck-billed platypus and the echidna) and marsupials. It has been suspected that the epipubics serve to support the pouch in marsupials like kangaroos and opossums—except that, problematically, both sexes of marsupials, pouched females and pouchless males, have these epipubic bones. Moreover, some female marsupials with epipubic bones entirely lack pouches.

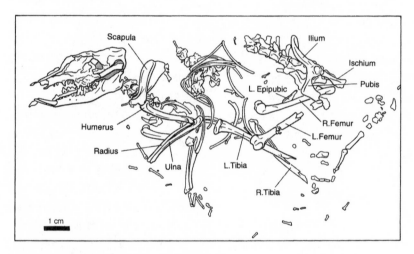

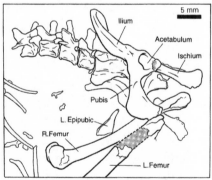

Skeletal elements of the Gobi Cretaceous mammal *Zalambdalestes*, and (in detail) the epipubic bones

In pondering this problem back at the museum, we turned up a paper that had been written on the measurements of epipubics in numerous species of living marsupials. The study showed that although both males and females have epipubics, these bones were always smaller in males, suggesting their lack of significance in supporting the abdominal region. We reasoned that epipubics were useful in strengthening the abdominal region in animals where the young are born at a very early stage and must be suspended outside the mother, either in a pouch or not (even pouchless female marsupials had to keep their flanks taut with all those young hanging from their teats), for a fairly long period of time. In contrast, the presence of these rigid struts would be disadvantageous for placental mammals, where the prolonged internal development of the young would distend the abdomen of the pregnant mother. Mammals like *Ukhaatherium* and *Zalambdalestes* had skull and tooth features very like more modern placental mammals rather than marsupials, though they also had epipubic bones. In a paper published in 1997, we proposed that these Gobi forms represented an intermediate stage in the evolution of placentals, where the redesign of the reproductive system lagged behind that of the skull and the teeth.

Some other mammal fossils from Ukhaa Tolgod also had a strong bearing on the history of more modern mammals. Since the 1920s the sharp-toothed deltatheridians, named for the triangular outline of their molars, had eluded proper placement within the mammalia. Some recent scholars suggested they were very primitive marsupials, while others contested this because of conflicting features of the dentition. Until we found several key specimens of these animals at Ukhaa Tolgod during the 1996–99 field seasons, this matter of classification was unsettled. Study of our beautifully preserved "deltas" resolved the issue. These creatures, about half the size of opossums, were indeed very much like modern marsupials. Instead of the placental number of three molars they had the marsupial number of four. They also had other marsupial markings—a broad, inwardly directed flange at the base of the jaw for the attachment of the jaw-closing muscles, a small splint of bone in the snout directed back toward the upper canine, and a particular pattern of canals for blood vessels in the middle-ear region.

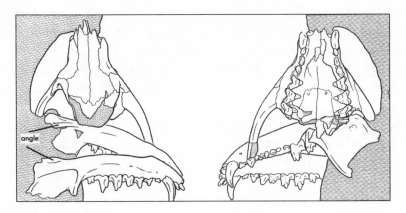

Top view (left) and ventral view (right) of the skull, palate, and jaws of *Deltatheridium*

Most interestingly, the Ukhaa Tolgod "deltas" showed the marsupial manner of tooth replacement. In placental mammals, all the milk premolars are replaced by permanent premolars, but marsupials and *Deltatheridium* replace only the milk premolar at the third position. The durability and permanence of the other premolars at an early stage are thought to relate to the very early need of the young marsupial to suckle at the nipple. Except for its well-developed mouth, tongue, and milk teeth, a newborn marsupial is so poorly developed that it looks like a worm or a grub with tiny limbs, closed eyes, and little bumps for ears; it is simply a suckling machine, critically dependent on those milk teeth to cling to the teat. *Deltatheridium* had tooth, skull, and skeletal features that suggested a marsupial form, function, and mode of life. We proposed that it represented a basal branch leading to living marsupials—a pattern that suggested the origins of marsupial forerunners in Asia and later radiations of more advanced marsupials in South America and Australia.

These anatomical minutiae and intricate evolutionary problems did not distract us from more general questions about Ukhaa Tolgod and other Gobi sites. Such questions are the same as the ones we are most frequently asked by nonspecialists. How did these animals live? How did they die? How were they buried? Several aspects of the fossil concentrations gave us some clues to the answers. First, there was the perfect condition of many of the fossils themselves—little disarticulation or abrasion, and thus no signs of scavenging and breakage, so common in vertebrate fossils. These remains had not been exposed on the ground surface for very long. Second, many of them took on an unusual pose, a "caught in the act" or "frozen in time" pose. These were buried remains of nesting oviraptorids, congregating ankylosaurs, familial mammals, and nests with hatchlings that had been emerging and feeding. Likewise, the fighting dinosaurs from Tugrigeen were a striking example of two animals killed and buried in the act. The Gobi sites preserved an extraordinary 80-million-year-old Pompeii of an ancient Cretaceous civilization.

It was also evident, though, that these deaths and burials were not part of one big catastrophe but rather of numerous catastrophes. The death assemblages—indicated by the various names of sublocalities like First Strike, Xanadu, Ankylosaur Flats, and Death Row—occurred at different levels in the rock section and therefore at different times in the sequence. How much time? It is not clear. The disasters could have been separated by thousands of years or only by a season. What we did know is they occurred over and over again.

So what killed these Cretaceous creatures? Since the expeditions of the American Museum in the 1920s it had been suggested that the animals were suddenly assaulted by sandstorms and suffocated and covered in great masses of sand. I myself had been persuaded of this scenario, given the detailed studies of the rocks and fossils of Bayan Mandahu conducted by geologists on the Sino-Canadian team. With the intention of doing our own autopsy of the victims at Ukhaa Tolgod, we invited a quiet, methodical geologist, David Loope from the University of Nebraska, to work with our team geologist, Lowell Dingus. Dave was an internationally recognized expert on sand dunes, sand formation, and sand sedimentation—a world-renowned sand man. In the 1996 and 1997 seasons he resolutely paced the outcrops for clues. Early on, he made a very interesting observation. There were two types of sand formations at Ukhaa Tolgod and other similar Gobi localities: one with impressive streaks or crossbeds that represented ancient windblown, migrating sand dunes, and the other with little in the way of crossbedding at all, which, Dave explained, were stabilized dunes, dunes that had settled in. The second type were no longer cross-bedded because they were once encrusted with plant life. Roots, wormholes, mammals' burrows, and other products of an active ecosystem had chewed up the layering of the sand (the fancy word for this phenomenon is bioturbation).

Then Dave surprised us with another observation. As hard as he tried, he could not find any vertebrate fossil bone preserved in the crossbedded sands. The only vertebrate fossils of any kind found in these ancient dunes were actually footprints. We had never seen di-nosaur footprints in these Gobi localities, but Dave demonstrated that we had walked right past them. At one magnificent facade of crossbedding, he took a knife and carved into a layer that was in-terrupted by some odd craterlike depressions. "Footprints, maybe *Protoceratops*, walking up the slope of a dune," he laconically submit-ted.

Dave challenged us to find fossil skeletons in these crossbedded sands, and we tried hard to do so. After some days, I became con-vinced that he was right. All those beautiful dinosaurs, lizards, and mammals were in the bioturbated sediments—in other words, in the stabilized, plant-covered dunes or in the small gullies between big migrating dunes. Dave then took stock of another clue. At first inspection, the skeletons seemed to be deposited in massive amounts of loose sand, but Dave observed streamers of tiny pebbles festooning the fossils. This suggested that water, or water mixed with sand and clay to form mud, was involved in the burial process. Instead of the sandstorm-catastrophe hypothesis, Dave and Lowell opted for another sequence of events. At various intervals, 80 mil-lion years ago, huge rainstorms caused mudflows and mud ava-lanches off the high dunes into the gullies where dinosaurs nested, ate, lived, died, and were buried.

This explanation offers an elegant fit between observations and scene-staging, but it doesn't resolve all the mysteries. It is possible, for instance, that many of the Gobi creatures died from starvation or disease, causes not easily decipherable in the fossil record. Much of the Gobi in the Cretaceous seems to have been a desert of dune fields, an inhospitable, arid place. Outcrops representing these an-

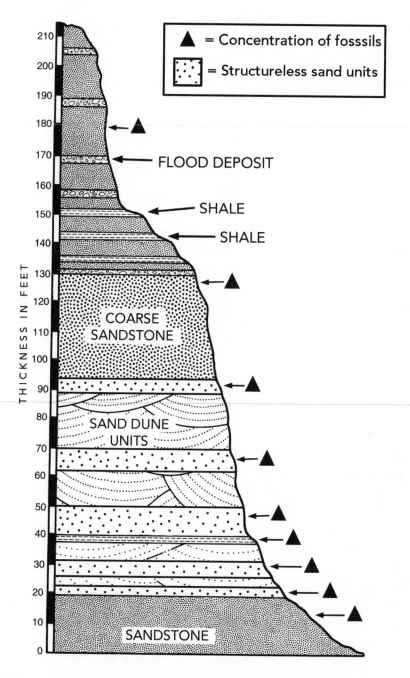

Geologic section at Ukhaa Tolgod, showing the levels where mass fossil concentrations were found

cient dune fields were completely barren of fossils, as Dave demonstrated. The good sites, like Ukhaa Tolgod and the Flaming Cliffs, concentrated the remains of animals that may have come to feed and to breed in comparatively sheltered areas. These animals thrived in oases in the middle of a desert—oases that were not always oases, not always secure from elements like rain, floods, and mudflows. The harsh oscillations of drought and flood, starvation and bounty, pestilence and health on the Serengeti or the Okavango Delta in Africa are a reenactment of the timeless tension between survival and catastrophe that is integral to the evolutionary history of life on earth.

The detective work, the excavation, and the prospecting at Ukhaa Tolgod continue to demand our attention all these years after the initial discovery of the site. Like the Flaming Cliffs, Ukhaa Tolgod will attract many future generations of paleontologists. Unfortunately, it has now also attracted egg thieves and bone poachers of the human variety. Mongolia has very strict national laws about the unauthorized exportation of fossils, but this has not deterred some wily criminals. During the last few field seasons we have noticed signs of irresponsible plunder at some of the more accessible sites like the Flaming Cliffs and Tugrigeen—hasty excavations surrounded by needlessly shattered bone. What new dinosaurs and other creatures important to science are being lost because of this greed? To make matters worse, dinosaurs and other fossils have demanded a high market price. When a *Tyrannosaurus* sells at auction for several million dollars, one can well imagine that a nesting *Oviraptor* could also fetch a handsome sum. This demand could even make legitimate excavations a dangerous business in an atmosphere of aggression far more serious than those of the Cope and Marsh dinosaur wars in nineteenth-century North America. The well-known dinosaur hunter Jack Horner advised me to start toting a gun or posting an armed guard at our locality. Ukhaa Tolgod is relatively remote and hard to locate, but we have no delusions that our precious site is safe from commercial or tourist raids. Talks with Mongolian authorities about establishing more controlled visitor sites at places like the Flaming Cliffs have been stalled because there are so many other pressing needs for resources and infrastructure in this impoverished nation. Sadly, the situation in Mongolia is simply a localized example of the worldwide abuse found in the fossil market. It is outrageous, for example, that the United States itself exercises one of the world's slackest sets of regulations concerning the removal of fossils from the country.

Greed drives market demand and inspires fossil stealers. It also inspires other villains. In 1995 I was visited in New York by a delegation of Mongolians who claimed to be officers of a newly formed Roy Chapman Andrews Society. Earlier, at their request and the recommendation of a friend, I had written a letter on my official stationery to the U.S. Information Service to help them gain short-term visas. The party arrived in my office with gifts and a proclamation to make me the first president of the society. They then asked me to sign a long decree written in old Mongolian script, an entreaty I politely refused. I thought nothing of this brief encounter until several months later. About to leave my office on a Saturday on my way to a Christmas party, I heard the fax machine start clacking. I was surprised to see a communication from Mongolia, a letter from the Ministry of Finance requesting the $5 million I had promised them! After a meeting the next Monday with the museum's general counsel, I wrote to my friend Donald Johnson, then the skilled, devoted U. S. ambassador to Mongolia. Meanwhile the Ministry tried to support the validity of its request with a fax copy of a letter sent to them. Sure enough, the letter contained a statement by me promising $5 million of the American Museum's money to the Roy Chapman Andrews Society. I noticed, however, that my letterhead and the signature looked like a paste-up job. Moreover, the text had several misspelled words, including my last name under my signature!

Ambassador Johnson soon e-mailed me the explanation. Apparently this contrivance of a society had used my letter to the U.S. Information Service as a template for a forged document. They then brought the forged letter to the Ministry of Finance and duped the officials there out of 90 million tugriks (the equivalent of $450,000) for a "development fee" to make a film about the life of Roy Chapman Andrews. Of course the Ministry was confident that, according to "my" letter, they would be receiving $5 million for their investment. This had also caused consternation among the executives of the Mongolian Academy of Sciences, who wondered why I had never bestowed such a reward on them for all their collaborative efforts. The forgers were apprehended and the forged letter with my signature appeared on the front page of the main Mongolian papers. Unfortunately, some still suspected I might be in cahoots with the forgers. The whole thing was a big mess.

Even during the subsequent 1996 field season, this ludicrous event had repercussions. The ringleader of the forgery scheme was now locked away in some frightful Mongolian jail, but the money had not been relocated. At Ukhaa Tolgod, we were visited by the mayor of the southern Gobi town of Gurvan Tes, one of the most isolated human outposts in the world, with a population not much larger than our field camp. But bad news travels fast. The mayor, upon meeting me, exclaimed in guttural Mongolian, "You are Novacek? The famous Novacek of the forgery scandal? Pleased to meet you!" Then, fortunately, he laughed.

Mongolia is indeed a country transformed by the rising prices of fossils and art, by the temptation of money schemes eager to scarf up the almighty dollar, and by the infusion of 30,000 tourists a year from Korea, Japan, the United States, and Europe. Yet it maintains some stubborn resistance to change. There is still only one major paved road in the country. Escape from the Gobi Desert at the end of a field season continues to be difficult. During our 1996 departure, our gas tanker slipped into a clay pit and started sinking. We dug at the metal beast for three days in the rain, and salvage was accomplished only through the most desperate of maneuvers: a metal cable the girth of an anaconda was wrapped around the fuel-laden tank, and the truck was wrenched up on only two wheels by two huge Russian trucks. One spark from that cable against the metal would have created a major new landform, a sooty crater, in the northern Gobi. Instead of scattering, though, we got under the truck and built a rock mound under its airborne wheels. More dangerous towing with taut cables finally brought the tanker to dry land.

I was so tired at the end of the last Gobi season that I slept solidly for two days in the Genghis Khan Hotel. When I awoke, I could hardly move my arms, paralyzed and aching from three days of pushing stubborn gearshifts and wrestling with the steering wheel. Those last couple of weeks in Ulan Bator are always miserable. We slog dirty equipment in the rain and try to cut through piles of red tape to get our dear fossils through customs and to the transport agencies, knowing well that after a period of preparation, illustration, and research they will eventually return to the country that produced them. As some of my college mates would say, "It's a lot of trouble for a few old bones." But what sensational bones! And what a sensational experience!

25

. . .

Last Chance Canyon

Cerro Condor is a hill in the middle of another beautiful nowhere. It rises from a dead river, an arroyo strewn with cobbles and sediments put there from some intermittent wall of water commuting from the high Andes, over the Patagonian plain, to the Atlantic. The rocks of Cerro Condor are tawny, chunky layers of mudstone, thrust upward in a heavenly urge and abruptly truncated at a summit only a few hundred feet above the arroyo. From the top one can look east to the slender ridges that girdle the Chubut drainage and directly below that to a cluster of deserted stone houses embraced by a stand of erect, shimmering poplars. A soft breeze cut with coolness from the Andes makes the short grass of Cerro Condor sing, and the light of the austral sun dances with the racing clouds, creating ripples of burnished yellow from summit to summit.

In November 1999, two days before the North American Thanksgiving, I sat on top of Cerro Condor breaking rocks. There were only a handful of others doing the same—my friend and colleague Guillermo Rougier; Tom Rich, a scientist trained at the American Museum in New York who had many years ago migrated to Australia; and some affable comrades of Guillermo's from the impressive new natural history museum in Trelew, capital of Chubut Province, Argentina. Guillermo knew this landscape well. By the time he'd started a postdoctoral fellowship with me at the museum a few years before, he had already found or collected some of Argentina's most spectacular dinosaurs and mammals. These included the big predator *Carnotaurus*, an animal distinguished from the carnivorous lookalike *Allosaurus* by a short, stubby horn protruding from the tip of its snout. Guillermo had also found sauropods as big as or bigger than any dinosaurs anywhere. Perhaps most spectacular was his find of *Amargasaurus*, a bizarre sauropod with elongate spines extending from the neck and trunk vertebrae. But now he, like me, was looking for small stuff.

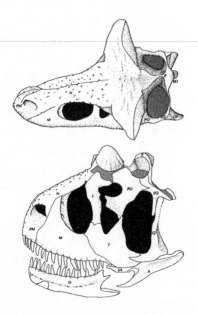

The skull of *Carnotaurus*, top and side views

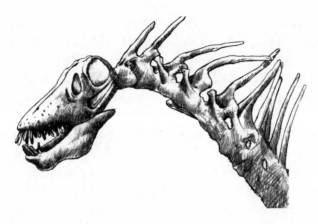

The skull and part of the skeleton of *Amargasaurus*

Tom Rich had spent a number of years chopping away at Cerro Condor and scrambling over the cliffs and buttes of the vast Chubut drainage. He too had been looking for small things in addition to dinosaurs. He wanted those mammals that might demonstrate affinities with species he had uncovered in the Cretaceous of Australia. Throughout most of the Mesozoic—the age of the dinosaurs—Australia and South America were broadly connected through Antarctica. Instead of an empire of foreboding ice, Antarctica at that time was lush and warm, a good land connection between two continents now displaced to either side of the Southern Hemisphere. At Tom's bidding we quarried briefly at a small outcrop of Jurassic sandstone near the top. Then we started picking up chunks and throwing them into sacks. "This stuff will screen-wash at the museum, maybe I'll get a tooth or two," he said. Tom had spent much time and money in this part of the world with little success. Now his funds had run out. This was probably his last season in Chubut, and he knew we were waiting in line. "Good luck. You guys have done well elsewhere, so maybe you'll strike it rich here," he said generously.

Both Guillermo and I were very busy in Mongolia, but Chubut offered new temptations. The Mesozoic and Early Tertiary rocks of South America are lavishly exposed in Chubut, exceeding even some of the expansive badlands of western North America and the Gobi. Surely earlier parties might have missed some great spots.

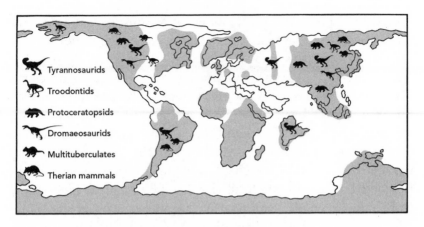

Continental relations during the Late Cretaceous, showing important
distributions of dinosaurs and mammals

On this bet, I raised some funding from the Antorchas Fundación
for an international collaboration that involved fieldwork, research,
and travel support and training for young Argentine scientists and
technicians. It was a welcome diversion from Mongolia. The Patagonian landscape was wild and grand, but the roads were well
maintained, the wine superb, and the range-fed beef extraordinary.
We had reached Cerro Condor from Trelew in three hours, traveling a distance that in Mongolia would have taken more than a day.
Yet, unlike the North American West, the countryside was still outback. There were no sprawling truck stops or fast-food establishments; there were only stone buildings in poplar groves and pickup
trucks loaded with alfalfa and ranch dogs.

We finished our work in only a couple of hours and started back
down the steep hill in the late afternoon light, laden with sacks full
of Jurassic rocks. This was the end of our brief frolic, our reconnaissance in Patagonia. In two days I would be waiting for a cab,
probably in a suffocating rain, at a grimy New York airport. As I
stepped into the truck I caught myself waving goodbye to an inert
mountain. I felt an emotional reunion with Patagonia, and I did
not want to leave it again. Somewhere to the west was a jagged
core of an old volcano called Pico Sur, where more than a decade
ago an undependable horse had almost done me in, but I had come
back for another look.

And we have more than sentimental reasons for wandering the canyons of Chubut Province. In addition to finding important fossils from Argentina's Jurassic, Cretaceous, and Early Tertiary, we were trying to plug some important gaps in the geologic record. The Cretaceous extinction event that erased the nonbird dinosaurs was well recorded, even at the end of the twentieth century, only for land vertebrates in western North America. For more than a decade we had searched energetically for that Cretaceous-Tertiary transition in the Gobi. But the 25-million-year gap between fossil-rich Late Cretaceous and fossil-rich middle Paleocene rocks in the Gobi was annoyingly persistent. We've got to plug that gap somewhere. The only way to determine whether the sudden extinction of dinosaurs and certain other land vertebrates was local or global is to find another transition that flanked the Cretaceous-Tertiary boundary, particularly a transition in rocks on another continent.

On the way back from Cerro Condor, I said to Guillermo, "You know, we should look in Antarctica, too. Most of the work there is confined to Seymour Island in the Weddell Sea, but I know there are great rocks on the Antarctic Peninsula. I've seen them. I can get the money and make the arrangements."

Guillermo gave a weary sigh, and then laughed. "Oh, Mike, I know you can, but please, not another expedition yet."

I shrugged my shoulders. I was sulking. "If we don't look there, someone else will. I think we should get in there soon."

Guillermo did not reply. He was looking out the window at the rainbow canyons of Chubut, probably thinking about his new house in Louisville and the anticipated arrival of his first child.

When dinosaur hunters made war in the late 1800s over fossil territories, much of nature was still largely unexplored, mysterious, even intimidating. But the golden age of exploration exploited that unfamiliar terrain. Many great accomplishments were driven by the noble quest for geographic and scientific knowledge. At the same time, many of these explorers were testing their own limits of endurance and seeking validation of the human capacity for heroism, hard work, and achievement. Andrews's Central Asiatic Expeditions in the 1920s had these diverse motives, and it would be hypocritical to decry them. Yet already during those early years of the twentieth century, the world was undeniably shrinking and becoming less challenging and wondrously elusive. Antarctica itself, discovered barely more than a century before, had been penetrated and "polarized" by humans dragged by domesticated dogs. By the time Mount Everest, the highest peak on earth, was successfully climbed in the 1950s, the last frontiers were to be found in sequestered deserts, mountains, and ice fields instead of on entire continents or islands. Only the oceans seemed cavernous and impenetrable on a scale comparable to the land during earlier centuries of human discovery. And eventually even the marine realm succumbed to the resolute curiosity of humans. Sonar and paleomagnetic technology, gravity-anomaly measurements, and other means used in unmanned surveys of the ocean produced extraordinary refined profiles of the ocean floor.

This lack of frontiers has recently prompted numerous rather contrived feats of daredevil travel. This new activity—some call it explornography—blurs the possibility of true and meaningful discovery. In a recent issue of *The New Yorker* devoted to both the good and bad sides of current exploration, the most gripping tale was a solo traverse of Central Park at midnight. Some exploits in faraway places seem merely gratuitous and stupid—not very different from that recent ascent 20,000 feet above the L.A. Basin accomplished by a terrified man in a lawn chair tied to weather balloons. I bristle when I read about modern adventurers intending to cross the Gobi in three days by motorcycle stocked with only power bars and some bottles of Gatorade. We've rescued a few of these despondent heroes, stranded by a flooded river. Exploration, even for its own sake, can still be compelling and exciting, but it takes real creativity to identify an arresting way to conquer some corner of the world.

There is, of course, still a very powerful motive—a desire for knowledge, whether of something as big as a dinosaur or as small as a bacterial cell in a thermal pool in Yellowstone. As valiant explorers were straddling the globe in the early twentieth century, it was clear that a little scientific knowledge was as empowering as, in an earlier epoch, the determination of latitude and longitude had been. One of the greatest planned exploits of all time, Ernest

Shackleton's 1914 attempt to cross the Antarctic continent from shore to shore, was thwarted by poor knowledge of Antarctic weather and the physics of ice floes. Although Shackleton saved his marooned crew in an astounding feat of bravery and altruism, his expedition's original goal was not attained. By the time someone actually accomplished it and crossed the icy continent, in 1958, the feat no longer had much meaning. The mystery of Antarctica had diminished and the world had experienced an infusion of unprecedented scientific knowledge. When in 1912 Scott reached the South Pole and subsequently perished with his men on the Ross Ice Shelf, Niels Bohr was already describing the structure of the atom. When in 1916 Shackleton piloted the escape of his men from the ice-choked Weddell Sea, Einstein was elaborating the general theory of relativity. Before the century was out scientists would discover plate tectonics, harness nuclear power, reach the moon, describe the universe in binary code, map all 30,000 genes of the human genome—and find a lot of fossils.

This reflection on the new age of scientific exploration brings me back to my own bailiwick. Instead of rearranging microchips I bang on often-penurious rocks for fossil bone, a pursuit I have loved from an early age. But what does paleontology really matter to anyone anyway? For one thing, the fossil record can be a powerful mirror reflecting back to us our evolutionary future. There are sobering lessons to learn from the past. The possibility of mass extinction, of the sudden, rampant destruction of large hunks of the biota, and the implications of this for the fickle quality of life such as we know it would not be readily evident without paleontologists and their fossil record. Moreover, that record has suggested important and relevant patterns. A recent scientific paper has shown that there may be a ten-million-year lag in the replenishment of ecosystems following a mass extinction event. This should dampen any misinformed optimism, however often published, about the supposed resilience of life.

The distant events of the fossil record are not wholly instructive about our current loss of biodiversity. This sweeping destruction of the living world in our time rivals in scope and impact the previous five great extinction events over the last 600 million years. Many scientists refer to this current biodiversity crisis as the sixth mass extinction. The earlier mass extinctions are puzzling in their genesis and variable in their effects, but it doesn't take much scientific sophistication to identify the cause of the current mass extinction. Expanding populations of humans are building structures, burning fuel, eating food, cultivating land, transporting goods—doing the things they have to do to survive—and they are marginalizing the natural world in the process. Unfortunately, this inexorable rise of human activity is accompanied by wholly gratuitous overfishing, overhunting, pollution, and poor land use. At the current rate of environmental degradation we seem to be losing approximately 30,000 species a year. The biodiversity of the planet could well be reduced by half by 2050. Meanwhile a very small cohort of scientists struggle desperately to document the vanishing biota, intent on discovering new species, describing their wondrous biological ways, and demonstrating their vital roles in the maintenance of ecosystems and their importance to human health, welfare, and quality of life. The fossil record is not needed to tell us what's going on—we can see extinction all around us. Paleontology should not be the only biological science of the future—the science of a dead planet.

Such momentous issues are admittedly distracting—as they should be—at a time when new scientific findings and recommendations should promote a greater sanity in government, policy makers, and society, where the stewardship of the living planet resides. I now spend a lot of nonresearch time helping to cultivate the infrastructure for biodiversity science at the American Museum where I work, in the United States, and abroad. I do believe that those of us lucky enough to have gained a reasonable sense of the legacy, diversity, and necessity of the biological world should contribute something to the effort to save it.

Meanwhile there are those bones in the lab—from Mongolia, Chile, Baja California, Wyoming, New Mexico, and elsewhere. In Chubut Province, Guillermo has even found recently a large sample of primitive mammals from the Cretaceous and an extremely significant little mammal jaw from the Jurassic quarry at Cerro Condor. But these precious pieces of the past represent only a small part of what's been collected over the centuries and only a few steps in the grandest experiment of all—the 3.5-billion-year evolutionary history of life on earth. When I was a boy I wanted to know all of that history. When I became a man I became excited with the notion that there was still so much to know.

This is a personal attachment to rocks and bones. It's allowed me to see extraordinary places and people in a world that, even during my few decades of field exploration, has become less exotic and more intimately entwined. But paleontology is not for everyone, and I respect the interests and talents in others who have no need of it. One day, many years ago, as I toiled at an outcrop almost too hot to touch in the heat of a New Mexico sun, Rodríguez, the lone cowboy of the Jornada del Muerto, rode up to me. He leaned over and squinted intently, trying to ascertain what I was up to. I lifted a broken, crummy spine of a 280-million-year-old pelycosaur to the shadow of his face. I tried to explain as best I could the nature of these ancient dragons and their importance to Dr. Vaughn and, by osmosis, to me. Rodríguez looked baffled, then simply bemused. He let out a low groan, his own version of a charitable laugh. Then he rode on.

Acknowledgments

. . .

The idea for this book came from people whose opinions I greatly value. Some years ago I wrote a nontechnical book about our discoveries of dinosaurs and ancient mammals in Mongolia's Gobi Desert. In addition to relating the trials, triumphs, and science integral to the Gobi expeditions, I inserted into that book some recollections of my fieldwork done when I was a naïve youth. During this fledgling phase, it was just as important to learn how to drive a Jeep, read a map, and walk the desert without getting lost, sunscorched, or snake-bitten as it was to find a good fossil. Many readers commented that these passages were among their favorite parts of the book because they dealt with visceral human feelings—ignorance and intimidation, but also pleasure and awe—that come with any nascent effort to do something well.

I liked those passages too. I decided to look back in more depth to those early years and also reflect on the expeditions to various places—the vacant lots of Los Angeles, the American West, Mexico, South America, and the Arabian Peninsula—that eventually brought me to the foot of the Flaming Cliffs in Mongolia. In *Time Traveler* I have also extended this account to include more recent Gobi expeditions and our latest foray into the windswept fossil terrain of Patagonia. These recollections are meant to show how a childhood of dinosaur dreams was transformed into a paleontological career. Although some of these stories deal with the adventure and triumph of discovery, they also relate the awkward struggle to learn something about field exploration, science, and, ultimately, oneself. On the safe premise that learning is continual and unbounded, I end the book with some emerging thoughts about why the fossil record tells us something about our evolutionary future.

In addition to the readers who inspired this work and the teachers, including my parents, to whom the book is dedicated, other people, knowingly or unknowingly, offered insights that influenced my writing. These include Richard Ellis, John Flynn, Charles Gallenkamp, Myles Gordon, Nancy Hechinger, Jen Meng, Craig Morris, Steven Novacek, Maureen O'Leary, Guillermo Rougier, and Ian Tattersall.

Time Traveler is a book not about a life, but largely about a life in the field. Fieldwork in paleontology as in other forms of exploration is usually not done solo. It is impossible to list all those who were helpful and inspiring companions to me over three decades of such work. Some of them were important as both expedition leaders and teachers, and then there are other friends and colleagues with whom I have shared many field seasons. These include Luis Chiappe, Jim Clark, Amy Davidson, Lowell Dingus, Ines Horowitz, Malcolm McKenna, Mark Norell, and Andy Wyss. In recent seasons I have enjoyed working with a new generation of bone hunters: Julia Clarke, Nick Frankfurt, Bolortsetseg Minchin, Bonnie Gulas, Pete Makovicky, and Diego Pol.

Field exploration to far-flung places cannot exist without international collaboration. I owe special thanks to Ishmael Ferrusquia and Oscar Carrenza in Mexico; Carlos de Smet, Daniel Frazinetti, and Renaldo Charrier in Chile; His Excellency Ali Jabr Alawi in Yemen; and Demberylin Dashzeveg, Rinchen Barsbold, Mongolian

Academy Secretary Galbaatar, and Academy President Chaadra, in addition to the many Mongolian expedition members, in Mongolia.

For our new expeditions to Patagonian Argentina, which were launched in 1999, I thank Bob Glynn, the Lampadia Foundation, and the Antorchas Fundación.

In recent years, my time outside paleontology has been well occupied by administrative duties at the American Museum of Natural History. Fortunately, there is pleasure in the latter, largely due to our inspiring Museum president, Ellen Futter, and my colleagues in the President's Council. I continue to be proud of representing the superb scientific staff of the Museum, who have upheld and even enhanced the leadership of the American Museum's science for over a century.

I once again thank Al Zuckerman, my agent, for his belief in my work and for the persuasive effect of his belief on others. For assistance in aspects of the book's preparation I thank Barbara Werscheck, Ed Heck, Lyn Merrill, Suzann Liberman, Jessica Bailly, and my daughter, Julie Novacek (who led me to the far-flung corners of the Web). Finally, the book was helped to become a reality because my editor, Elisabeth Sifton, has the uncanny capacity of blending insight, enthusiasm, and an affable and powerful encouragement to get on with it.

M.N.
New York
August 2001

Notes

■ ■ ■

In addition to citations from the literature, the notes refer to Web site sources. Many Web sites dealing with dinosaurs and related scientific subjects are overly speculative, are poorly documented, and lack critical review, however. The scientific Web sites included here are deemed to be reasonably informative, accurate, and authoritative.

I. DINOSAUR DREAMER

4 "...Vasquez Rocks, had been degraded to a roadside rest area...." This is now Vasquez Rocks County Park, thirty miles north of Los Angeles, a site that was used to film episode 19, "The Arena," in the television series *Star Trek*. See *http://www.geocities.com/ Hollywood/Theater/1991/st arena.html*. A brief description and photograph of the park is provided by the Boy Scouts: *http://www.geocities.com/troop484/camping/ vasquez.html*.

4 "I liked books, ones on dinosaurs...." My favorite book about dinosaurs when I was seven was Andrews, R. C., 1953, *All About Dinosaurs*, Random House, New York.

7 "But in the late 1950s, neither I nor paleontological experts...." The renaissance in dinosaur paleontology began with observations that *Archaeopteryx*, the ancient Jurassic bird preserved with feathers, looked very much like certain small carnivorous dinosaurs called coelurosaurs. Although such resemblance had been hinted at in much earlier work, it was first clearly emphasized by John Ostrom, who reviewed his contribution as well as earlier interpretations in Ostrom, J. H., 1974, "*Archaeopteryx* and the origin of

flight," *Quarterly Review of Biology* 49: 27–47. A key paper that places this connection in a modern evolutionary context is Gauthier, J., 1986, "Saurischian monophyly and the origin of birds," in Padian, K. (ed.), *The Origins of Birds and the Evolution of Flight*, Memoirs, California Academy of Sciences, 8, pp. 1–55. A massive literature, some credible, some not, claims that dinosaurs by analogy with birds were active, warm-blooded creatures of high intelligence that probably exhibited parental care in addition to other aspects of a complex repertoire of behavior. I review this work in Novacek, M. J., 1996, *Dinosaurs of the Flaming Cliffs*, Anchor/Doubleday, New York. For overviews see Dingus, L., and Rowe, T., 1998, *The Mistaken Extinction: Dinosaur Evolution and the Origin of Birds*, Freeman, New York.

7 "Today a few scientists still object to this connection. . . ." Detractors argue that dinosaurs are not clearly related to birds and that allegedly birdlike dinosaurs occur too late in the fossil record to be related to earlier forms like *Archaeopteryx*. Many of these studies are vague in stating what dinosaurs *are* closely related to, usually opting for some linkage with an ancient reptilian archosaur group that also may have branched into crocodilians. A comprehensive but inconsistent argument against the connection between bird and dinosaurs is Feduccia, A., 1996, *The Origin and Evolution of Birds*, Yale University Press, New Haven, CT. Meanwhile, the accumulating data on dinosaur anatomy, as represented by their eggs, embryos, and adults, concretely demonstrates their close affinity with birds, as discussed by Norell, M. A., 1995, "Origins of the feathered nest," *Natural History* 104 (6): 58–61.

7 "This includes newly discovered fossils from China. . . ." Spectacular remains from the Late Jurassic to Early Cretaceous Liaoning quarries in northern China clearly show evidence of feather impressions in animals that otherwise resemble small, nonflying theropod dinosaurs. These feathers were obviously not used for flight, but may have functioned as insulation and in display just as in living, warm-blooded birds. Key studies of the Liaoning fossils are Ji, Q., Currie, P. J., Ji, S.-A., and Norell, M. A., 1998, "Two feathered theropods from the Upper Jurassic/Lower Cretaceous strata of northeastern China," *Nature* 393: 753–61; and Ji, Q., Norell, M. A., Gao, K.-Q., Ji, A.-A., and Ren, D., 2001, "The distribution of integumentary structures in a feathered dinosaur," *Nature* 410: 1084–88.

8 "I read about real field expeditions. . . ." Early field expeditions to these areas, discussed in more detail in later chapters, are reviewed in Colbert, E. H., 1968, *Men and Dinosaurs*, Dutton, New York.

8 "The tar gives the park its Spanish name, La Brea." A general description is in Harris, J. M., and Jefferson, G. T. (eds.), 1985, *Rancho La Brea: Treasures of the Tar Pits*, Natural History Museum of Los Angeles County; see also *http://www.tarpits.org/exhibits/fossils/main.html* and *http://www.ucmp.berkeley.edu/quaternary/labrea.html*.

9 ". . . the Los Angeles County Natural History Museum. . . ." See *http://www.nhm.org*.

9 ". . . a mural by the great painter of prehistory, Charles R. Knight." A very similar mural is displayed at the American Museum of Natural History in New York, but I didn't discover this until much later. Knight, C. R., 1947, *Animal Drawing: Anatomy and Action for Artists*, McGraw-Hill, New York. A retrospective of Knight's works is in Czerkas, S. M., and Glut, D. F., 1982, *Dinosaurs, Mammoths and Cavemen: The Art of Charles R. Knight*, Dutton, New York.

10 "Horses native to North America went extinct around La Brea time. . . ." Modern summaries of horse evolution, including their extinction in North America, include MacFadden, B. J., 1988, "Horses, the fossil record, and evolution: a current perspective," *Evolutionary Biology* 22: 131–58; and MacFadden, B. J., 1992, *Fossil Horses: Systematics, Paleobiology, and Evolution of the Family Equidae*, Cambridge University Press, Cambridge, England.

11 ". . . Time-Life book *The World We Live In*." Barret, L., and Life staff (eds.), 1955, *The World We Live In*, Time Inc., New York.

2. THE CANYON OF TIME

14 "... road cuts of bronze sandstone. ..." The thick section of Miocene marine rocks exposed in the Santa Monica Mountains and adjacent regions, known as the Topanga Formation, is briefly described in *Southern California Geology* by Ruth Lebow, at *http://www.lalc.k12.ca.us/target/fragile habitats/geo of Ca.html*; and Sharp, R. P., et al., 1993, *Geology Underfoot in Southern California*, Mountain Press Publishing, Missoula, MT.

15 "... an outrageous gash in the earth, the Grand Canyon." Baars, D. L., Buchanan, R. C., and Charlton, J. R., 1994, *The Canyon Revisited: A Rephotography of the Grand Canyon, 1923/1991*, University of Utah Press, Salt Lake City; Redfern, R., 1980, *Corridors of Time: 1,700,000,000 Years of Earth at Grand Canyon*, Times Books, New York; and Billingsley, G., 2001, "The Erosion of the Grand Canyon," in Mathez, E. A. (ed.), 2001, *Earth: Inside and Out*, American Museum of Natural History and New Press, New York, pp. 117–23. See also *http://www.kaibab.org/geology/gc geol.htm*.

16 "... born about 20 million years ago. ..." Mathez, *Earth: Inside and Out*."

16 "... it merely chronicles an end-zone event. ..." Ibid.

16 "They were wrought by big forces. ..." Ibid.

17 "Give or take a few more layers. ..." Ibid.

18 "Where are the younger rocks?" A good current survey of the regional geology of the rock formations around the Grand Canyon in northern Arizona and southern Utah is Baars, D. L., 2000, *The Colorado Plateau: A Geologic History*, University of New Mexico Press, Albuquerque.

19 "There are ancient rocks in Greenland. ..." Mathez, *Earth: Inside and Out*, pp. 16–48; Condie, K. C. (ed.), 1994, *Archean Crustal Evolution*, Elsevier, Amsterdam/New York.

19 "Taking account of this entire timescale. ..." An engaging, highly readable review of the formulation of the earth calendar is Berry, W. B. N., 1968, *Growth of a Prehistoric Time Scale*, Freeman, San Francisco.

19 "Even Leonardo da Vinci ... recorded layers of ancient sea rocks. ..." The codex, which contains Leonardo's notes on fossils and strata, is explained in Farago, C. (synopsis and commentary), Pedretti, C. (translation), 1996, *Codex Leicester: A Masterpiece of Science/ Leonardo da Vinci*, American Museum of Natural History, New York.

19 "... Nicholas Steno refined Leonardo's perspective. ..." Berry, *Growth of a Prehistoric Time Scale*.

19 "Others, like the nineteenth-century British geologist Charles Lyell. ..." Ibid.

19 "It was Lyell who planted the appreciation of this vast timescape in Charles Darwin." Lyell's great work *Principles of Geology* was a favorite of Darwin's and always close at hand during the voyage of the *Beagle*. Milner, R., 1993, *The Encyclopedia of Evolution*, Henry Holt, New York, pp. 286–87.

21 "Tiny traces of radioactive elements. ..." Technical explanations of radiometric dating are many, descriptions for a general readership few. Two are Benton, M., 1993, "Life and time," in Gould, S. J. (ed.), 1993, *The Book of Life*, Norton, New York, pp. 22–37. See also my *Dinosaurs of the Flaming Cliffs*.

21 "A combination of radiometric dating. ..." Berry, *Growth of a Prehistoric Time Scale*; Harland, W. B., Armstrong, R. L., Cox, A. V., Craig, L. E., Smith, A. G., and Smith, D. G., 1990, *A Geologic Time Scale, 1989 Edition*, Cambridge University Press, Cambridge, England.

22 "After all, the human lineage, represented by *Australopithicus* from Africa. ..." Tattersall, I., 1995, *The Fossil Trail*, Oxford University Press, Oxford, England.

22 "Remote time, or deep time, may be beyond our ken." This point is emphatically made in Gee, H., 1999, *In Search of Deep Time: Beyond the Fossil Record to a New History of Life*, Free Press, Simon and Schuster, New York.

22   "... the 1.7 billion years it does preserve is time enough." Redfern, *Corridors of Time*.
22   "The mountains at the beginning of this history...." Ibid.

3. EARLY EXPEDITIONS

26   "... Wisconsin was mainly a land of very old, boring rocks...." An excellent geologic map of the state is provided by the Wisconsin Geological and Natural History Survey at *http://www.uwex.edu/wgnhs/bdrk.htm*.
26   "... representing the last great retreat of the glaciers." The last ice age, when the ice sheets advanced between 50,000 and 15,000 years ago, is in fact called the Wisconsin Glacial Interval. Harland et al., *Geologic Time Scale*.
26   "... the Green River meandering through a great canyon...." Lageson, D. R., and Spearing, D. R., 1988, *Roadside Geology of Wyoming*, Mountain Press Publishing, Missoula, MT, pp. 69–74.
27   "These hollows exposed old marine rocks...." The Early Paleozoic geology of Wisconsin is represented on the Wisconsin Geological and Natural History Survey map, *http://www.uwex.edu/wgnhs/bdrk.htm*. See also the survey map of glacial deposits: *http://www.uwex.edu/wgnhs/iceage.htm*.
27   "The fossil was indeed a relative of pill bugs...." Arthropods are earth's most diverse organisms of any kind (but bacteria, fungi, and various other groups have hardly been surveyed), there being about 1.5 million named living species. A summary of the diversity of different biological groups is provided in Raven, P. H., "What have we lost, what are we losing?" in Novacek, M. J. (ed.), 2001, *The Biodiversity Crisis: Losing What Counts*, American Museum of Natural History and New Press, New York, pp. 58–61.
27   "It was a trilobite...." This particular trilobite shall be forever unknown to me, but a good candidate is some species of *Phacops*. For a discussion see Eldredge, N., 1985, *Time Frames*, Simon and Schuster, New York, pp. 57–71.
28   "... the Field Museum of Natural History...." Now called simply the Field Museum. See *http://www.fmnh.org*.

4. THE PLATEAU OF THE DRAGONS

36   " 'I plan to make a last stab at the Abo and Yeso formations....' These formations are exposed west of the Sacramento Mountains in southern New Mexico. The central reference is Vaughn, P. P., 1969, "Early Permian vertebrates from southern New Mexico and their paleozoogeographic significance," *Natural History Museum of Los Angeles County Contributions in Science*, no. 166: 1–22.
36   "... Caballo Mountains." Good regional coverage of this range and nearby features can be found in Chronic, H., 1987, *Roadside Geology of New Mexico*, Mountain Press Publishing, Missoula, MT. A more technical account of the relevant paleontology is in Vaughn, "Early Permian vertebrates."
36   "... Monument Valley...." Baars, *Colorado Plateau*.
36   "... amphibians ..." A detailed survey of vertebrate fossils is given in Carroll, R. L., 1988, *Vertebrate Paleontology and Evolution*, Freeman, New York.
36   " 'I feel a bit of a traitor....' " Ibid.
36   "Some Permian creatures were big...." Ibid.
36   "Pennsylvanian-aged creatures were less imposing." Ibid.
36   "Many of these Pennsylvanian amphibians...." Ibid.
37   "Others, like nectridians...." Ibid.
38   "*Dimetrodon* was a dominant predator...." Ibid.
38   "Some lizards use a webbing...." Ibid.
38   "*Dimetrodon* did not glide...." Ibid.

38 "Heat was a Permian thing. . . ." Vaughn, "Early Permian vertebrates."

38 "Permian rocks in places like Utah. . . ." Ibid.

39 "*Dimetrodon* is a dramatic animal. . . ." Here I have tried to summarize the basic relationships of the major groups of vertebrates, called amniotes, that are more committed to land life than amphibians. The amniotes contain the extant groups birds, mammals, crocodiles, lizards, snakes, turtles, and their extinct relatives—dinosaurs, synapsids, pterosaurs, icthyosaurs, and more. I have given more detailed information on the phylogeny of the amniotes in *Dinosaurs of the Flaming Cliffs*; see also Dingus and Rowe, *Mistaken Extinction*.

39 "As odd as it may seem. . . ." Carroll, *Vertebrate Paleontology*.

41 ". . . they call the whole mess the Colorado Plateau . . ." Baars, *Colorado Plateau*. There is an excellent Web site on the geological history of the southwest United States and related topics offered by Dr. Rob Blakely: *http://vishnu.glg.nau.edu/rcb/RCB.html*.

43 "The movements of the crustal plates. . . ." The revolution in earth sciences is the subject of a massive literature. Good general references are Miller, R., 1983, *Continents in Collision*, Time-Life Books, Alexandria, VA; and Mathez, *Earth: Inside and Out*. Classic volumes containing a good sampling of some of the original technical and semitechnical articles are Cloud, P. (ed.), 1970, *Adventures in Earth History*, Freeman, San Francisco; and Cox, A. (ed.), 1973, *Plate Tectonics and Geomagnetic Reversals*, Freeman, San Francisco.

44 ". . . but such upheavals are also constructive. . . ." Mathez, *Earth: Inside and Out*.

44 "There are some controversial theories. . . ." A comprehensive but thoroughly disputed volume is Carey, S. W, 1976, *The Expanding Earth*, Elsevier, Amsterdam.

44 "The key to answering the question. . . ." Miller, *Continents in Collision*; and Cox, *Plate Tectonics*.

45 ". . . 1.7-billion-year-old rocks at the bottom of the Grand Canyon. . . ." Redfern, *Corridors of Time*.

45 ". . . 3.5-billion-year-old ones in Greenland:" Dalrymple, G. B., 1991, *The Age of the Earth*, Stanford University Press, Stanford, CA. See also U.S. Geological Survey Web site on geologic time: *http://pubs.usgs.gov/gip/geotime/age.html*.

45 "It was subsequently found that. . . ." Miller, *Continents in Collision*.

45 "Plate boundaries that primarily involve trenches. . . ." Ibid.

46 ". . . deep-focus earthquakes." Cox, *Plate Tectonics*.

46 "The tensional forces. . . ." Ibid.

46 ". . . shallow-focus earthquakes." Ibid.

46 "Earthquakes are more likely to occur where the geological action is. . . ." Atwater, B. F., 2001, "Averting earthquake surprises in the Pacific Northwest," in Mathez, *Earth: Inside and Out*, pp. 89–93.

46 "It can explain the birth of many mountain ranges. . . ." Miller, *Continents in Collision*.

47 "The crust of the Colorado Plateau. . . ." Baars, *Colorado Plateau*.

47 ". . . 'roof of the world'. . . ." This titanic collision was recognized in earlier work reviewed in Hurley, P. M., 1968, "The confirmation of continental drift," *Scientific American* 218 (4): 52–64. A recent technical study that relates the collision event and changing Asian geology to global patterns in climate is Zhisheng, A., Kutzbach, J. E., Prell, W. L., and Porter, S. C., 2001, "Evolution of Asian monsoons and phased uplift of the Himalayan-Tibetan plateau since Late Miocene times," *Nature* 411: 62–66.

47 "Geologists recognize a number of factors. . . ." Baars, *Colorado Plateau*.

47 "The Colorado Plateau remains a lofty. . . ." Ibid.

47 "Fossils need quiet beds. . . ." The general principles of fossil deposition and preservation are discussed in Benton, "Life and Time"; and Novacek, *Dinosaurs of the Flaming Cliffs*.

48 "Instead they emerge from the rock through the cooperation of wind and water. . . ." Novacek, *Dinosaurs of the Flaming Cliffs*.

49 "Many paleontologists spend an inordinate amount of time. . . ." Ibid.
49 ". . . *Tseajaia campi.* . . ." Moss, J. L., 1972, "The morphology and phylogenetic relation-
    ships of the Lower Permian tetrapod *Tseajai campi* Vaughn (Amphibia: Seymouriamor-
    pha)," *University of California Publications in Geological Sciences* 98: 1–72.
49 ". . . Organ Rock Shale. . . ." Ibid.

CHAPTER 5. THE ROOKIES

54 "We spied a field of great sand dunes. . . ." These are part of the extensive dunes known
    as the Imperial Sand Dunes, a BLM Recreation and Wilderness Site that also served as
    a site for a sequence in the movie *Star Wars.* A description of the natural history of the
    Imperial Sand Dunes is given at *http://www.desertusa.com/sandhills/sandhillsorg.html.*
55 " 'Basin and range,' he said. . . ." A classic treatment of the geology of the basins and
    ranges of the western United States is McPhee, J., 1980, *Basin and Range,* Farrar, Straus
    & Giroux, New York.
55 " 'Everything important in vertebrate evolution. . . .' " The general outlines of the fossil
    record of vertebrates are reviewed in Carroll, *Vertebrate Paleontology*; and Novacek, *Di-
    nosaurs of the Flaming Cliffs.*
56 "From there we drove north and east. . . ." At the time of our expedition, vertebrate fos-
    sils had been found in four canyons, the Tularosa, the Domingo, the Labrocita, and the
    Cottonwood. These canyons dissected exposures of the Yeso, Labrocita, and Abo forma-
    tions. The fossil localities were largely restricted to the Abo Formation. A detailed de-
    scription is given in Vaughn, "Early Permian vertebrates."

6. JOURNEY OF DEATH

57 ". . . .a blue-white fragment of a pelycosaur spine. . . ." Fossils from the Abo Formation
    localities are described in Vaughn, "Early Permian vertebrates." A general description of
    New Mexico geology is provided in Chronic, *Roadside Geology of New Mexico.*
58 ". . . the village of Perm. . . ." Berry, *Growth of a Prehistoric Time Scale,* pp. 91–95.
62 ". . . we found only little flecks. . . ." Vaughn himself made the first discovery of
    Dimetrodon in the Caballo Mountains. Vaughn, "Early Permian vertebrates."
65 ". . . like Mars, where any evidence of life. . . ." Mathez, *Earth: Inside and Out,* pp. 196–99.

7. THE PALEONTOLOGICAL CHAIN GANG

67 ". . . lowly objects, called coprolites. . . ." The study of fossil feces is time-honored, ex-
    tending back to works like Buckland, W., 1823, *Reliquiae Diluvianae,* 1978 reprint,
    J. Murray, London, and Arno Press, New York; and Buckland, W., 1835, "On the discov-
    ery of coprolites, or fossil faeces, in the Lias at Lyme Regis, and in other formations,"
    Transactions of the Geological Society of London, series 2, 3 (1): 223–36. A good modern
    summary is Chin, K., 1997, "Coprolites," in Currie, P. J., and Padian, K. (eds.), 1997, *En-
    cyclopedia of Dinosaurs,* Academic Press, San Diego, pp. 147–51. The article is also avail-
    able at *http://darwin.apnet.com/dinosaur/chin.htm.*
67 ". . . digested prey items, like small Permian amphibians." Chin, "Coprolites."
67 "Nonetheless, there is a small cohort of coprolite specialists. . . ." Ibid.
67 ". . . both coprolites and fossil eggs containing embryos can be CAT-scanned." A central
    facility for CAT-scanning vertebrate fossils is managed by Timothy Rowe, University of
    Texas. A relevant article, accompanying an available CD, is Rowe, T., Carlson, W., and
    Bottorff, W., 1993, *Thrinaxodon: Digital Atlas of the Skull,* University of Texas Press
    (CD-ROM). Information can be found at *http://www.ucmp.berkeley.edu/synapsids/
    rowe/rowe.html.*
69 "We stopped at Ghost Ranch. . . ." Colbert, E. H., 1995, *The Little Dinosaurs of Ghost*

Ranch, Columbia University Press, New York; and Schwartz, H. L., and Gillette, D. D., 1994, "Geology and taphonomy of the *Coelophysis* quarry, Upper Triassic Chinle Formation, Ghost Ranch, New Mexico," *Journal of Paleontology* 68: 1118–30.

69 "This form, *Coelophysis*. . . ." Schwartz and Gillette, "Geology and taphonomy."

69 ". . . these innocuous-looking creatures were cannibals." Ibid.; also Norell, M. A., Gaffney, E. S., and Dingus, L., 1995, *Discovering Dinosaurs*, Knopf, New York.

70 ". . . a great and terrible event. . . ." A comprehensive general reference is Erwin, D. H., 1993, *The Great Paleozoic Crisis*, Columbia University Press, New York. See also Gould, *Book of Life*.

70 ". . . an ugly, rather stumpy beast. . . ." Carroll, *Vertebrate Paleontology*.

71 "A Permian catastrophe. . . ." Erwin, D. H., 2001, "Lessons from the past: biotic recoveries from mass extinctions," *Proceedings of the National Academy of Sciences* 98: 5399–5403.

72 "It is the greatest of the five great extinctions. . . ." Ibid.

72 ". . . the phase of decimation somewhere around 100,000 years. . . ." This has been suggested by some scientists, as reviewed in Hoffmann, H. L., 2000, *National Geographic* 198: 100–13.

72 "Another limitation of our understanding. . . ." Ibid.

73 "It took several million years for these communities to reemerge. . . ." Erwin, "Lessons from the past"; and Kirchner, J. W., and Weil, A., 2000, "Delayed biological recovery from extinctions throughout the fossil record," *Nature* 404: 177–80.

74 " 'Central Texas,' he said. . . ." The classic papers by Romer on this subject include Romer, A. S., 1928, "Vertebrate faunal horizons in the Texas Permo-Carboniferous red beds," *University of Texas Bulletin*, no. 2801: 67–108. Other authors and publications include Olson, E. C., 1958, "Fauna of the Vale and Choza, 14. Summary, review, and integration of the geology and the faunas," *Fieldiana, Geology* 10: 307–448; and Berman, D. S., 1970, "Vertebrate fossils of the Lueders Formation, lower Permian of north-central Texas," *University of California Publications in Geological Sciences* 86: 1–39.

76 "This ritual inaugurated. . . ." Techniques are described in many sources, including Novacek, *Dinosaurs of the Flaming Cliffs*.

76 ". . . the animals in this quarry near the Sangre de Cristos. . . ." Vaughn, P. P., 1969, "Upper Pennsylvanian vertebrates from the Sangre de Cristo Formation of Central Colorado," *Natural History Museum of Los Angeles County Contributions in Science*, no. 164: 1–28.

77 "The quarry did contain bone shards. . . ." Ibid.

77 "They were important fossils. . . ." Carroll, *Vertebrate Paleontology*.

77 ". . . fossils have much to tell us. . . ." Novacek, M. J., 1992, "Fossils, topologies, missing data, and the higher level phylogeny of eutherian mammals," *Systematic Biology* 41: 58–73; Gauthier, J., Kluge, A. G., and Rowe, T., 1988, "Amniote phylogeny and the importance of fossils," *Cladistics* 4: 105–209; Donoghue, M., Doyle, J., Gauthier, J., Kluge, A., and Rowe, T., 1989, "The importance of fossils in phylogeny reconstruction," *Annual Reviews of Ecology and Systematics* 20: 431–60; and Novacek, M. J., and Wheeler, Q. C. (eds.), 1992, *Extinction and Phylogeny*, Columbia University Press, New York.

77 ". . . fossils of *Archaeopteryx*. . . ." Ostrom, "*Archaeopteryx* and the origin of flight"; Wellnhofer, P., 1988, "A new specimen of the Archaeopteryx," *Science* 240: 1–55.

78 ". . . the Pennsylvanian Period, which predates the Permian." Harland et al., *Geologic Time Scale*.

79 ". . . a world of steaming bogs and swamps. . . ." Schopf, J. W. (ed.), 1992, *Major Events in the History of Life*, Jones and Bartlett, Boston, pp. i–xv, 190; Gould, *Book of Life*.

79 "It was the only known assemblage. . . ." Vaughn, "Upper Pennsylvanian vertebrates."

79 ". . . an unpublished master's thesis on the stratigraphy of the area. . . ." The author of the thesis was not Pierce but K. G. Brill, Jr., who also published Brill, K. G., Jr., 1952,

"Stratigraphy in the Permo-Pennsylvanian zeugeosyncline of Colorado and northern New Mexico," *Geological Society of America Bulletin* 49: 2041–75.

79 ". . . the undramatic name 'Interval 300 Quarry'. . . . " Vaughn, "Upper Pennsylvanian vertebrates."

80 "Although only a few plant fossils. . . ." Ibid.

81 "These boulders also had. . ." These are very primitive jawless fishes characteristic of the earlier Paleozoic. Carroll, *Vertebrate Paleontology.*

8. RED ROCKS

84 ". . . a brilliant red sequence of Pennsylvanian, Permian, and Triassic rocks. . . ." Baars, *Colorado Plateau.*

84 ". . . amphibian skeletons like *Tseajaia*. . . ." Moss, "Morphology and phylogenetic relationships."

84 ". . . the lower and older Halgaito Shale. . . ." Vaughn, P. P., 1962, "Vertebrates from the Halgaito tongue of the Cutler Formation, Permian of San Juan County, Utah," *Journal of Paleontology* 36: 529–39; Vaughn, P. P., 1964, "Vertebrates of the Organ Rock Shale of the Cutler Group, Permian of Monument Valley and vicinity, Utah and Arizona," *Journal of Paleontology* 38: 567–83.

85 "But his evasiveness belied. . . ." Vaughn, P. P., 1966, "Comparison of the Early Permian Vertebrate Faunas of the Four Corners region and north-central Texas," *Natural History Museum of Los Angeles County Contributions in Science*, no. 105: 1–13.

86 "Vaughn's famous professor. . . ." Romer, A. S., 1958, "The Texas Permian redbeds and their vertebrate fauna," in *Studies on Fossil Vertebrates presented to David Meredith Seares Watson*, University of London, Athlone Press, pp. 157–79.

87 "In a paper published in 1966. . . ." "Comparison of the Early Permian." Contrary to Vaughn's thesis and prediction, *Dimetrodon* was more recently found in the upland Permian facies of north-central New Mexico, as noted by Berman, D. S., 1977, "A new species of *Dimetrodon* (Reptilia, Pelycosauria) from a non-deltaic facies in the Lower Permian of north-central New Mexico," *Journal of Paleontology* 51: 108–15.

9. BACK TO THE BONES

90 "We climbed Mount Ellen. . . ." The controversial history of wilderness protection in the region around Mount Ellen is reviewed in "SUWA Winter, 1997, How the BLM 'Sanitized' Mt. Ellen," at *http://www.suwa.org/images/newsletters/1997winter_mt_ellen.html.*

92 "The skeleton belonged to a new and important amphibian. . . ." Vaughn, P. P., 1972, "More vertebrates, including a new microsaur, from the Upper Pennsylvanian of Central Colorado," *Natural History Museum of Los Angeles County Contributions in Science*, no. 223: 1–30.

93 ". . . an eccentric meteorologist named Alfred Wegener." His famous and controversial work was Wegener, A., 1912, "Die Entstehung der Kontinente," *Geologishe Rundschau* 3: 276–92 (reprinted as Wegener, A., 1966, *The Origins of Continents and Oceans*, Dover, New York). The controversy is reviewed in Miller, *Continents in Collision.*

93 "The refutation of some theories. . . ." Vaughn, "Comparison of the Early Permian"; Berman, "New species of *Dimetrodon.*"

10. THE PITS

97 "By 1970 the tar pits had been excavated. . . ." An updated history is Harris, J. M., and Jefferson, G. T. (eds.), 1985, *Rancho La Brea: Treasures of the Tar Pits*, Natural History Museum of Los Angeles County. Also see the excellent Web sites *http://www.tarpits.org/ exhibits/fossils/main.html* and *http://www.ucmp.berkeley.edu/quaternary/labrea.html.*

97   "... the existence of human remains...." Merriam, J. C., 1914, "Preliminary report on the discovery of human remains in an asphalt deposit at Rancho La Brea," *Science* 40: 198–203.

97   "But the bones of La Brea woman...." See Harris and Jefferson, *Rancho La Brea*, as well as relevant Web sites.

97   "Since the loss of carbon 14 increases over time at a constant rate...." The classic paper on the subject is Libby, W. F., 1961, "Radiocarbon dating," *Science* 133: 621–29.

98   "One of the most common of these species is *Canis dirus*, the dire wolf." Harris and Jefferson, *Rancho La Brea*.

98   "The scientific name here is in two parts...." This binomial naming system is called the Linnaean system after the great eighteenth-century taxonomist Carolus Linnaeus. Linnaeus, C., 1735, *Systema Naturae, sive Regna tria Naturae systematice proposita per Classes Ordines, Genera & Species*, Fol. Lugduni Batavorum. A traditional treatment of systematics and taxonomy is in Mayr, E., 1942, *Systematics and the Origin of Species*, Columbia University Press, New York. More modern treatments are Hennig, W., 1966, *Phylogenetic Systematics*, University of Illinois Press, Urbana; Eldredge, N., and Cracraft, J., 1980, *Phylogenetic Patterns and the Evolutionary Process*, Columbia University Press, New York; Nelson, G., and Platnick, N. I., 1981, *Systematics and the Biogeography—Cladistics and Vicariance*, Columbia University Press, New York; Queiroz, K. de, and Donoghue, M. J., 1988, "Phylogenetic systematics and the species problem," *Cladistics* 4: 317–38; Nixon, K. C., and Wheeler, Q. D., 1992, "Extinction and the origin of species," in Novacek, M. J., and Wheeler, Q. D. (eds.), *Extinction and Phylogeny*, Columbia University Press, New York, pp. 119–43; Gaffney, G., Dingus, L., and Smith, M., 1995, "Why cladistics?" *Natural History* 104 (6): 33–35.

99   "Some bones are better for this purpose than others." The study of bone preservation and comparative durability during fossilization is called taphonomy, a term coined in the classic paper Efremov, J. A., 1940, "Taphonomy, a new branch of paleontology," *Pan-American Geologist* 74: 81–93.

100   "... La Brea lion, *Panthera atrox*, and the saber-toothed cat...." These are dominant species of the La Brea fauna. See Harris and Jefferson, *Rancho La Brea*.

100   "Between July 1913 and September 1915, the Los Angeles County Natural History Museum crews had made ninety-six numbered excavations." This work is summarized in Marcus, L. F., 1960, "A census of the abundant large Pleistocene mammals from Rancho La Brea," *Natural History Museum of Los Angeles County Contributions in Science*, no. 38: 1–11.

100   "In a 1960 paper, Leslie Marcus reported on a census of the best pits...." Ibid.

102   "In our more modern excavation...." Since the early 1970s the main pit where we worked has been subject to continuous excavation, as related at *http://www.tarpits.org./pit91/pit91001.html*. It is now thought that the pits represent an age span of between 3,000 and 9,000 years. Recent notable work on the La Brea excavations and collections includes Biknevicius, A., Van Valkenburgh, B., and Walker, J., 1996, "Incisor size and shape: implications for feeding behaviors in saber-toothed cats," *Journal of Vertebrate Paleontology* 16: 510–21. This study showed that the carnivores at La Brea suffered much more tooth breakage and damage than carnivores living on the African savanna today, suggesting that *Canis dirus*, *Smilodon*, and *Panthera atrox* were forced to consume kills more fully, chewing the bones and thereby injuring the teeth. A more general account of the scenario is given by Van Valkenburgh, B., 1994, "Tough times in the tar pits," *Natural History* 103 (4): 84–85.

II. THE AGE OF MAMMALS REVISITED

105   "... 'beautiful Mission Valley.' " Stock, C., 1938, "A tarsiid primate and a mixodectid from the Poway Eocene, California," *National Academy of Sciences Proceedings* 24: 288–93.

106 "... a decent slice of time called the later Eocene epoch. ..." The significance of the San Diego mammal fauna for understanding the later Eocene on a broader scale was explained by Black, C. C., and Dawson, M. R., 1966, "A review of the Eocene mammal faunas from North America," *American Journal of Science* 264: 321–49; and Lillegraven, J. A., 1973, "Terrestrial Eocene vertebrates from San Diego County," in Ross, A., and Dowlen, R. (eds.), *Studies on the Geology and Geologic Hazards of the Greater San Diego Area, California*, Guidebook for May Field Trip, San Diego Association of Geology and Engineering Geology, pp. 27–32.

106 "Mammals are indeed very old. ..." Kermack, D. M., and Kermack, K. A. (eds.), 1971, "Early mammals," *Linnaean Society Zoological Journal* 50, suppl. 1; Lillegraven, J. A., Kielan-Jaworowska, Z., and Clemens, W. A. (eds.), 1979, *Mesozoic Mammals: The First Two-Thirds of Mammalian History*, University of California Press, Berkeley; Szalay, F. S., Novacek, M. J., and McKenna, M. C. (eds.), 1993, *Mammal Phylogeny (Vol. 1): Mesozoic Differentiation, Multituberculates, Monotremes, Early Therians, and Marsupials*, Springer Verlag, New York; and Carroll, *Vertebrate Paleontology*.

106 "One unique feature of mammals is the presence of three tiny bones in the ear region. ..." Gaupp, E., 1913, "Die Reichertsche Therorie (Hammer-, Amboss- und Kieferfrage)," *Arch. f. Anat. u. Entwicklungsgeschichte*, 1912, pp. 1–416. The Gaupp theory of the evolution and development of the mammalian ear ossicles is a triumph of comparative biology. This is reviewed in Novacek, M. J., 1993, "Patterns of diversity in the mammalian skull," in Hanken, J., and Hall, B. K., *The Skull (Vol. 2): Patterns of Structural and Systematic Diversity*, University of Chicago Press, Chicago, pp. 438–534.

107 "Another aspect of the skeleton. ..." Szalay et al., *Mammal Phylogeny: Mesozoic Differentiation*.

108 "This more complex battery row of teeth in mammals is different in another important way. ..." Ibid.

109 "This difference between marsupials and placentals with respect to teeth. ..." Ibid. The early systematics and classification of the major groups of mammals is covered in Gregory, W. K., 1910, "The orders of mammals," *Bulletin of the American Museum of Natural History* 27: 1–524.

109 "The earliest mammals. ..." Novacek, M., 1994, "A pocketful of fossils," *Natural History* 103 (4): 40–43.

111 "At the end of the Cretaceous. ..." A few of the many notable publications on the Cretaceous extinction event are: Alvarez, L. W., 1983, "Experimental evidence that an asteroid impact led to the extinction of many species 65 million years ago," *Proceedings of the National Academy of Science* 80: 627–42; Florentin, J.-M., Maurrasse, R., and Sen, G., 1991, "Impacts, tsunamis, and the Haitian Cretaceous-Tertiary boundary layer," *Science* 252: 1690–93; Glen, W., 1990, "What killed the dinosaurs?" *American Scientist* 78 (4): 354–70; Hunter, J., 1994, "Lack of a high body count at the K-T boundary," *Journal of Paleontology* 68(5): 1158; Sheehan, P. M., Fatovsky, D. E., Hoffman, R. G., Berghaus, C. B., and Gabriel, D. L., 1991, "Sudden extinction of the dinosaurs: Late Cretaceous, Upper Great Plains, U.S.A.," *Science* 254: 835–39; Stanley, S. M., 1987, *Extinctions*, Scientific American Books, New York; Williams, M. E., 1994, "Catastrophic versus noncatastrophic extinction of the dinosaurs: testing, falsifiability, and the burden of proof," *Journal of Paleontology* 68: 183–90; Archibald, J. D., 1966, *Dinosaur Extinction at the End of an Era: What the Fossils Say*, Columbia University Press, New York; Novacek, *Dinosaurs of the Flaming Cliffs*; and Dingus and Rowe, *Mistaken Extinction*.

111 "The period following the Cretaceous and its extinction event is the Tertiary. ..." The Cretaceous-Tertiary history is extensively reviewed in Novacek, M. J., 1999, "100 million years of land vertebrate evolution: the Cretaceous-Early Tertiary Transition," *Annals*

of the Missouri Botanical Gardens 86: 230–58, and many references cited therein. A comprehensive compendium on Tertiary mammal evolution and classification is Janis, C. M., Scott, K. M., and Jacobs, L. L., 1998, *Evolution of Tertiary Mammals of North America (Vol. 1): Terrestrial Carnivores, Ungulates, and Ungulatelike Mammals,* Cambridge University Press, Cambridge, England.

111   "The Tertiary epochs. . . ." Harland et al., *Geologic Time Scale*; Berggren, W. A., Kent, D. V., Aubry, M.-P., and Hardenbol, J. (eds.), 1995, *Geochronology, Time Scales, and Global Stratigraphic Correlation,* Special Publications 54, Society for Sedimentary Geology, Tulsa, OK.

111   "The first of these Tertiary epochs, the Paleocene. . . ." Novacek, "100 million years."

112   "The Paleocene also experienced some global warming. . . ." Ibid. See also Janis, C. M., 1993, "Tertiary mammal evolution in the context of changing climates, vegetation, and tectonic events," *Annual Reviews of Ecology and Systematics* 24: 467–500; Wing, S. L., and Sues, H. D., 1992, "Mesozoic and Early Cenozoic terrestrial ecosystems," in Behrensmeyer, A. K., et al. (eds.), *Terrestrial Ecosystems Through Time,* University of Chicago Press, Chicago, pp. 327–416.

112   "In fact, the early Eocene was a time when the earth had perhaps the highest global average annual temperatures. . . ." Novacek, "100 million years."

112   "In the later Eocene. . . ." Ibid.

112   "This trend continued through the Oligocene and Miocene. . . ." Ibid. See also Prothero, D. R., and Schoch, R. M. (eds.), 1994, *Major Features of Evolution,* Publication of the Paleontological Society, no. 7.

12. ROAD CUTS AND FOSSIL VERMIN

114   "The next step was laundry duty." The screen-washing technique is described in McKenna, M. C., 1965, "Collecting microvertebrate fossils by washing and screening," in Kummel, B., and Raup, D., *Handbook of Paleontological Techniques,* Freeman, San Francisco, pp. 193–204.

115   ". . . notably for me, insectivorans." As far back as Gregory, "The order of mammals" (1910), insectivorans, because of their many primitive traits, were recognized as a key group to resolving the basic branches of the placental mammals.

117   " 'Study nature, not books.' " Agassiz's aphorism is recorded in the memoirs of David Star Jordan, 1922, *The Days of Man: Being Memories of a Naturalist, Teacher, and Minor Prophet of Democracy,* World Book Company, Yonkers-on-Hudson, New York. Jordan notes, "On the walls we put several mottos taken from Agassiz's talks to us: STUDY NATURE NOT BOOKS. . . ." (pp. 117–18).

118   ". . . until I had a fat manuscript. . . ." My master's thesis, submitted in 1973, was much later published in highly revised versions. Novacek, M. J., 1976, "Insectivora and Proteutheria of the later Eocene (Uintan) of San Diego County, California," *Natural History Museum of Los Angeles County Contributions in Science,* no. 283: 1–52; Novacek, M. J., 1985, "The Sespedectinae, a new subfamily of hedgehog-like insectivores," *American Museum Novitates,* no. 2833: 1–24; Novacek, M. J., and Lillegraven, J. A., 1979, "Terrestrial vertebrates from the later Eocene (Uintan) of San Diego County, California: A Conspectus," in Abbott, P. L. (ed.), *Eocene Depositional Systems,* Society of Economic Paleontology and Mineralogy, Pacific Section, Special Publication, pp. 69–79.

118   "The map was really in the form of a tree. . . ." The principles of tree building or phylogenetics are amply covered in Hennig, *Phylogenetic Systematics*; Eldredge and Cracraft, *Phylogenetic Patterns*; Nelson and Platnick, *Systematics and the Biogeography*; Novacek and Wheeler, *Extinction and Phylogeny.*

119   ". . . the published literature on fossil mammals was not very helpful in this tree-building procedure." These complaints were shared by many, as represented by Nelson and Plat-

nick, *Systematics and the Biogeography.* A recent compilation of more modern applications of phylogenetics is Gee, H. (ed.), 2000, *Shaking the Tree: Readings from Nature in the History of Life,* University of Chicago Press, Chicago.

119 ". . . George Gaylord Simpson. . . ." Simpson's rather "artful" and vague approach is described in Simpson, G. G., 1945, "The principles of classification and a classification of mammals," *Bulletin of the American Museum of Natural History* 85: 1–350.

120 "Dr. Percy Butler. . . ." A notable work by this insectivoran expert is Butler, P. M., 1972, "The problem of insectivore classification," in Joysey, K. A., and Kemp, T. S. (eds.), *Studies in Vertebrate Evolution,* Winchester Press, New York, pp. 253–65.

120 ". . . the American Museum of Natural History. . . ." See *http://research.amnh.org/vertpaleo.*

120 ". . . UC Berkeley. . . ." See *http://www.ucmp.berkeley.edu/index.html.*

121 "In Europe, a garbage pit near the German town of Messel. . . ." A beautifully illustrated volume is Koenigswald, W.V., and Storch, G., 1998, *Messel: ein Pompeii der Palaeontologie,* Jan Thorbecke Verlag GmbH, Sigmaringen, Germany.

122 ". . . thin-layered shale in northern China. . . ." Ji et al., "Two feathered theropods"; Ji et al., "Distribution of integumentary structures."

13. BADLANDS, BONES, AND BONE HUNTERS

126 "The hot spots for paleontologists. . . ." Good guides include Alt, D., and Hyndman, D. W., 1986, *Roadside Geology of Montana,* Mountain Press Publishing, Missoula, MT; Lageson, D. R., and Spearing, D. R., 1988, *Roadside Geology of Wyoming,* Mountain Press Publishing, Missoula, MT.

126 "The rocks record the gradual change. . . ." McPhee, J., 1986, *Rising from the Plains,* Farrar, Straus & Giroux, New York.

126 "These Jurassic dinosaur communities. . . ." Lageson and Spearing, *Roadside Geology of Wyoming.*

127 "The dinosaur-dominated communities. . . ." Ibid.

127 "Somewhat younger Tertiary rocks with lots of fossil mammals. . . ." Woodburne, M. O. (ed.), 1987, *Cenozoic Mammals of North America: Geochronology and Biostratigraphy,* University of California Press, Berkeley.

127 ". . . the basins are clearly identifiable. . . ." Lageson and Spearing, *Roadside Geology of Wyoming.*

128 ". . . the rock layers dip gently down toward the center of the basin. . . ." Ibid.

128 "The basins and badlands of Montana and Wyoming. . . ." Dingus and Rowe, *Mistaken Extinction.*

128 "In 1855 Dr. Ferdinand Vandiveer Hayden led an expedition to the Western territories. . . ." The history of fossil hunting in the basins and ranges of the Western states, including the bitter rivalry between Marsh and Cope, is vividly chronicled in Colbert, E. H., 1968, *Men and Dinosaurs,* Dutton, New York, and in many works cited therein. Norell, M. A., Gaffney, E. S., and Dingus, L., 1995, *Discovering Dinosaurs,* Knopf, New York, contains an updated treatment of historical events relating to the dinosaur collections at the American Museum of Natural History.

134 "Berkeley crews worked at the northern edges of the Washakie Basin. . . ." Lageson and Spearing, *Roadside Geology of Wyoming,* pp. 28, 60–61.

134 ". . . badlands of Hell Creek. . . ." Alt and Hyndman, *Roadside Geology of Montana,* pp. 363–64.

134 ". . . known as the Crazy Mountain Field. . . ." Simpson, G. G, 1937, "The Fort Union of the Crazy Mountain Field and its mammalian fauna," *U.S. National Museum Bulletin,* no. 169: 1–287.

135 ". . . now laid claim to the Cretaceous Lance Creek. . . ." Clemens, W. A., 1964, "Fossil mammals of the type Lance Formation, Wyoming, Pt. 1: Introduction and Multituberculata," *University of California Publications in Geological Sciences* 48: 1–105.

135 ". . . some choice Eocene localities west of the Wind River Mountains." West, R. M., 1970, "Sequence of mammalian faunas of Eocene age in the northern Green River Basin, Wyoming," *Journal of Paleontology* 14: 142–47.

135 ". . . in the Bighorn Basin. . . ." Gingerich, P. D., 1980, *Early Cenozoic Paleontology and Stratigraphy of the Bighorn Basin, Wyoming*, University of Michigan Papers in Paleontology, no. 24.

14. HELL CREEK IS FOR DINOSAURS

136 ". . . at Hell Creek. . . ." Alt and Hyndman, *Roadside Geology of Montana*, pp. 363–64.

137 "But these sites. . . ." Simpson, G. G., 1937, "The Fort Union of the Crazy Mountain Field and its mammalian fauna," *U.S. National Museum Bulletin*, no. 169: 1–287.

138 ". . . Don Russell in the Paris Basin. . . ." Russell, D. E., 1964, "Les Mammiferes Paleocenes d'Europe," *Museum Naturelle Histoire National Memoir*, n. s. ser. C. 13: 1–324.

140 ". . . the type, or reference, specimen of *Tyrannosaurus rex* . . . discovered here by Barnum Brown. . . ." Norell, Gaffney, and Dingus, *Discovering Dinosaurs*.

140 "The Hell Creek is also unique as a rock sequence." Archibald, J. D., 1996, *Dinosaur Extinction at the End of an Era: What the Fossils Say*, Columbia University Press, New York; Novacek, *Dinosaurs of the Flaming Cliffs*; Dingus and Rowe, *Mistaken Extinction*.

141 ". . . carbonized black coal, known as the Z coal." Archibald, *Dinosaur Extinction*.

141 "This so-called ten-foot gap. . . ." Ibid.

141 "Despite the mysterious gap. . . ." Ibid.

142 " 'Multi' is short for multituberculate. . . ." Novacek. "A pocketful of fossils."

143 "It was *Triceratops*. . . ." Dodson, P., and Currie, P. J., 1990, "Neoceratopsia," in Weishampel, D. B., Dodson, P., and Osmolska, H. (eds.), *The Dinosauria*, University of California Press, Berkeley, pp. 593–618; Norell, Gaffney, and Dingus, *Discovering Dinosaurs*; Sampson, S. D., 1995, "Horns, herds, and hierarchies," *Natural History* 104 (6): 36–40.

145 "We were now in the Bighorn Basin. . . ." Bown, T. M., 1980, "The Willwood Formation (Lower Eocene) of the southern Bighorn Basin, Wyoming, and its mammalian fauna," in Gingerich, *Early Cenozoic Paleontology*, pp. 127–38.

147 "This was a rodent skull. . . ." The skull belonged to the primitive genus *Paramys.*, described by Wood, A. E., 1962, "Early Tertiary rodents of the Family Paramyidae," *Transactions of the Philosophical Society*, new series, no. 52: 1–261.

148 "I even snagged the big mammal of the summer. . . ." Carroll, *Vertebrate Paleontology*, p. 531.

148 "Classic locations like Como Bluff. . . ." Colbert, *Men and Dinosaurs*.

148 "It was frustrating that. . . ." Andrews, R. C., 1932, *The New Conquest of Central Asia*, vol. 1 of *Natural History of Central Asia*, American Museum of Natural History; Buffetaut, E., and Leloeuff, J. L., 1994, "The discovery of dinosaur eggshells in nineteenth-century France," in Carpenter, K., Hirsch, K. F., and Horner, J. A. (eds.), *Dinosaur Eggs and Babies*, Cambridge University Press, Cambridge, England, pp. 31–34; Horner, J. R., 1982, "Evidence for the colonial nesting and 'site fidelity' among ornithischian dinosaurs," *Nature* 297: 675–76; Carpenter, K., 1982, "Baby dinosaurs from the Late Cretaceous Lance and Hell Creek formations and a description of a new species of theropod," *Contributions to Geology of the University of Wyoming* 20: 123–34.

149 ". . . searched all summer in the Evanston Overthrust. . . ." This is sometimes called the Overthrust Belt. Lageson and Spearing, *Roadside Geology of Wyoming*, pp. 191–207.

149 "There he found the bones of a big flying reptile. . . ." Lawson, D. A., 1975, "Pterosaur from the latest Cretaceous of West Texas: discovery of the largest flying creature," *Science* 187: 947–48.

150 ". . . the then largest known flying reptile, *Pteranodon*. . . ." For a review of various pterosaurs see Padian, K., 1997, "Pterosauria," in Currie and Padian, *Encyclopedia of Dinosaurs*, pp. 613–17.

150 "He named the animal *Quetzalcoatlus northrupii*. . . ." Lawson, "Pterosaur from the latest Cretaceous."

150 ". . . Jack Horner and his team found rich nesting grounds. . . ." Horner, "Evidence for the colonial nesting."

15. THE CURIOUS BEASTS OF OLD BAJA

153 ". . . very archaic mammals called leptictids. . . ." Novacek, M. J., 1986, "The skull of leptictid insectivorans and the higher-level classification of eutherian mammals," *Bulletin of the American Museum of Natural History* 183: 1–111.

155 ". . . La Mision locality. . . ." Savage, D. E., and Russell, D. E., 1983, *Mammalian Paleofaunas of the World*, Addison-Wesley, Reading, MA, p. 237.

156 "This was Vieja Baja, Old Baja of the Miocene epoch." Ibid.

157 ". . . Shark's Tooth Hill near Bakersfield. . . ." Ibid.

157 "From my vantage point I could see a pod of Miocene whales. . . ." Ibid.

157 ". . . a pack of long slinky hyenalike forms chasing a frail young camel." This scene can be confidently reconstructed from the well-known assemblages of Hemingfordian age within the Miocene. Ibid.

159 "Highway 1 cut through this river valley. . . ." Morris, W. J., 1967, "Baja California Late Cretaceous dinosaurs," *Science* 155: 1539–41.

159 ". . . chunky but impressive dinosaur bones." Ibid.

160 ". . . one small marsupial-like mammal called *Gallolestes*. . . ." Lillegraven, J. A., 1976, "A new genus of therian mammals from the Late Cretaceous 'El Gallo Formation,' Baja California, Mexico," *Journal of Paleontology* 30: 437–43. See also Lillegraven, J. A., 1972, "Preliminary report on Late Cretaceous mammals from the El Gallo Formation, Baja California, Baja Del Norte, Mexico," *Natural History Museum of Los Angeles County Contributions in Science*, no. 232: 1–11.

160 "In a paper published in *Science* in 1967. . . ." Morris, "Baja California Late Cretaceous dinosaurs."

161 "This form, which Molnar named *Labocania*. . . ." Molnar, R. F., 1974, "A distinctive theropod dinosaur from the Upper Cretaceous of Baja California (Mexico)," *Journal of Paleontology* 48: 1009–17.

162 "Here in 1966, crews from the Instituto. . . ." Morris, W. J., 1966, "Fossil mammals from Baja California: new evidence of Early Tertiary migrations," *Science* 153: 1376–78; Morris, W. J., 1968, "A new early Tertiary perissodactyl, *Hyracotherium seekensi*, from Baja California," *Natural History Museum of Los Angeles County Contributions in Science*, no. 151: 1–11.

162 "On the basis of this small sample, Bill Morris. . . ." Ibid.

162 "This seemed rather odd. . . ." Atwater, T., 1970, "Implications of plate tectonics for the Cenozoic tectonic evolution of western North America," *Geological Society of America Bulletin* 81: 3513–36.

163 "A more complete sample should either show. . . ." The large number of rich fossil localities in western North America allowed a reasonably clear distinction between late Paleocene and early Eocene land mammal ages. Rose, K. D., 1980, "Clarkforkian Land Mammal Age: revised definition, zonation, and tentative intercontinental correlation," *Science* 208: 744–46.

163 "Another kind of evidence could be highly useful too." We did eventually apply this technique, as described in Flynn, J. J., Cipolletti, R. M., and Novacek, M. J., 1989, "Chronology of early Eocene marine and terrestrial strata, Baja California, Mexico," *Geological Society of America Bulletin* 101: 1182–96.

16. BAKING BELOW THE BUTTES

167 "There is probably no lovelier cactus garden on earth." An excellent reference is Coyle, J., and Roberts, N. C., 1975, *A Field Guide to the Common and Interesting Plants of Baja California*, Natural History Publishing, La Jolla, CA.

167 "We had U.S.G.S. and Mexican topographical maps. . . ." Gastil, R. C., Phillips, R. P., and Allison, E. C., 1975, "Reconnaissance geology of the state of Baja California," *Geological Society of America Memoir* 140: 1–40. The Mexican map is Hoja Rosarito, HII, D69, 1:50,000; CETENAL (=INEGI), 1977.

168 ". . . Lomas Las Tetas de Cabra, or Hills of the Goat's Teats." Mexican topographical map, Hoja Rosarito, HII, D69.

168 ". . . Andy found the tooth of *Meniscotherium*. . . ." The mammals from the Lomas Las Tetas fauna were surveyed in Novacek, M. J., Flynn, J. J., Ferrusquia-Villafranca, I., and Cipolletti, R. M., 1987, "An early Eocene (Wasatchian) mammal fauna from Baja California," *National Geographic Research* 3: 376–88. A detailed monograph on the fauna is Novacek, M. J., Ferrusquia-Villafranca, I., Flynn, J. J., Wyss, A. R., and Norell, M., 1991, "Wasatchian (early Eocene) mammals and other vertebrates from Baja California, Mexico: The Lomas Las Tetas de Cabra Fauna," *Bulletin of the American Museum of Natural History* 208: 1–88.

168 ". . . that elusive primitive horse *Hyracotherium*. . . ." Novacek et al., "Wasatchian mammals."

169 ". . . a primitive herbivorous animal, probably *Hyopsodus*. . . ." Ibid.

169 "The specimens were not exceptionally well preserved. . . ." Ibid.

175 "We withstood the conditions at Las Tetas. . . ." Ibid.

176 "We manage slowly to accumulate an interesting array of fossils. . . ." Ibid.

176 "*Wyolestes* and its kin, the mesonychids. . . ." Ibid.

177 "Unfortunately, the kind of igneous rocks. . . ." Further explanation of the use of radiometric analysis is given above and in Benton, "Life and time"; and Novacek, *Dinosaurs of the Flaming Cliffs*.

178 "This direction is determined by an amazing physical phenomenon." The phenomenon of paleomagnetism and its use in the development of a paleomagnetic timescale are described in many articles in Cox, *Plate Tectonics*.

178 "One of the great discoveries in the study of the magnetic properties. . . ." Ibid.

178 "Periods of normal and reversed polarity. . . ." Ibid

179 "The magnetic signals, usually called paleomagnetic signals. . . ." Ibid.

179 ". . . John and Bob could focus on a time line that seemed most consistent with their paleomagnetic measurements. . . ." The results of this work are published in Flynn, Cipolletti, and Novacek, "Chronology of early Eocene strata."

179 "When they calculated for such movements. . . ." Ibid.

179 "Other geological evidence was consistent with Baja's having. . . ." Atwater, "Implications of plate tectonics."

180 "The age determination turned out to be decisive." Flynn, Cipolletti, and Novacek, "Chronology of early Eocene strata."

180 "With these new fossils, the Baja fauna looked if anything more like early Eocene faunas. . . ." Novacek et al., "An early Eocene mammal fauna"; Novacek et al., "Wasatchian mammals."

180 "Most notably, a fossil site of similar mammals in the high Arctic...." Hickey, L. J., West, R. M., Dawson, M. R., and Choi, D. K., 1983, "Arctic terrestrial biota: paleomagnetic evidence of age disparity with mid northern latitudes during the late Cretaceous and early Tertiary," *Science* 221: 1153–56.

181 "Our work on the Baja fauna was surprising evidence...." Novacek et al., "Wasatchian mammals."

181 "Indeed, the paleontological results jibed well...." Hickey et al., "Arctic terrestrial biota."

181 "... average annual global temperature during the early Eocene...." Novacek, "100 million years."

181 "... *Science* published a preliminary analysis...." Flynn, J. J., and Novacek, M. J., 1984, "Early Eocene vertebrates from Baja California: evidence for intercontinental age correlations," *Science* 224: 151–53.

181 "Several other publications...." Novacek et al., "An early Eocene mammal fauna"; Novacek et al., "Wasatchian mammals"; Flynn, Cipolletti, and Novacek, "Chronology of early Eocene strata."

17. WHALES ON MOUNTAINTOPS

182 "Stories were told...." Carpenter, Hirsch, and Horner, *Dinosaur Eggs and Babies*; Andrews, *New Conquest of Central Asia.*

183 "Our findings at Las Tetas de Cabra...." Novacek et al., "An early Eocene mammal fauna"; Novacek et al., "Wasatchian mammals"; Flynn, Cipolletti, and Novacek, "Chronology of early Eocene strata."

183 "... a well-known paleontologist, George Gaylord Simpson...." Simpson published many technical papers on the South American faunas. His field experiences are related in Simpson, G. G., 1934, *Attending Marvels: A Patagonian Journal*, University of Chicago Press, Chicago. His nontechnical review of the evolution of the South American mammal fauna is Simpson, G. G., 1980, *Splendid Isolation: The Curious History of South American Mammals*, Yale University Press, New Haven, CT.

184 " 'They are actually whale vertebrae....' " Carroll, *Vertebrate Paleontology.*

184 "... Henry Fairfield Osborn...." Osborn was both a prolific paleontologist and the imperious director of the American Museum of Natural History. Preston, D. J., 1986, *Dinosaurs in the Attic: An Excursion into the American Museum of Natural History*, St. Martin's Press, New York.

184 "... Edward Drinker Cope...." Colbert, *Men and Dinosaurs.*

185 "Fossil whales were indeed common in certain beds...." Whale fossils are abundant in certain Tertiary localities along the North American and South American coasts. Savage and Russell, *Mammalian Paleofaunas.*

185 "... the vertiginous rise of the Andes." Miller, *Continents in Collision.*

185 "Neighboring Argentina had all the fossils...." As reviewed in Simpson, G. G., 1984, *Discoverers of the Lost World*, Yale University Press, New Haven, CT.

185 "Not the least of these was Charles Darwin." Darwin assiduously recorded his observations in Darwin, C., 1839, 1952, *Journal of researches into the geology and natural history of the various countries visited by the H.M.S. Beagle*, facsimile reprint of the first edition, Hafner, New York.

186 "... local scientists like the great Ameghino brothers...." Ameghino, F., 1906, "Les formations sedimentaires du Cretace supérieur et du Tertiaire de Patagonie avec un parallèle entre leurs faunes mammalogiques et celles de l'ancien continent," *Anales del Museo Nacional Buenos Aires* 14: 1–568; Matthew, W. D., 1915, "Climate and Evolution," *Annals of the New York Academy of Sciences* 24: 171–318; Simpson, G. G., 1948, 1967, "The begin-

ning of the Age of Mammals in South America," *Bulletin of the American Museum of Natural History* 91: 1–232 and 137: 1–260.

186 "One of the most colorful, persistent, and successful of these early explorers was John Bell Hatcher. . . ." Hatcher, J. B., 1985, *Bone Hunters in Patagonia, Narrative of the Expedition*, Ox Bow Press, Woodbridge, CT (reprinted from Hatcher, J. B., 1903, *Princeton University Expedition to Patagonia, 1896–1899*, Princeton University).

187 " 'There were two physicians. . . .' " Ibid., pp. 92–93.

188 " 'The personnel, the party, the name of the ship and her officers. . . .' " Ibid., p. 101.

188 "Likewise, discoveries made in Brazil. . . ." Simpson, *Splendid Isolation*.

188 ". . . the great supercontinent of Pangea. . . ." Miller, *Continents in Collision*.

189 "By the Late Cretaceous, South America had become a gargantuan island." Ibid.

189 "There in the southern half of the world. . . ." Patterson, B., and Pascual, R., 1968, "The fossil mammal fauna of South America," *Quarterly Review of Biology* 43 (4): 409–51; Simpson, *Splendid Isolation*; Marshall, L. G., and Cifelli, R. L., 1989, "Analysis of changing diversity patterns in Cenozoic land mammal age faunas, South America," *Palaeovertebrata* 19: 169–210.

189 "There was a third curious radiation of mammals in South America." Simpson, *Splendid Isolation*.

190 ". . . two additional groups, the rodents and the primates." Ibid.

192 "Darwin himself pondered the demise. . . ." Darwin, *Journal of researches*, pp. 81–87.

192 ". . . the Great American Interchange, was actually more of a one-way massacre." Simpson, *Splendid Isolation*; Webb, S. D., 1976, "Mammalian faunal dynamics of the Great American Interchange," *Paleobiology* 2: 220–34; Stehli, F. G., and Webb, S. D. (eds.), 1985, *The Great American Biotic Interchange*, Plenum Press, New York; Webb, S. D., 1991, "Ecogeography and the Great American Interchange," *Paleobiology* 17: 266–80.

192 "Despite a rough balance in the exchange. . . ." Webb, "Mammalian faunal dynamics."

193 "Why such an uneven outcome. . . ." Ibid.

193 ". . . George Gaylord Simpson proposed that the northern species of placental mammals. . . ." Simpson, *Splendid Isolation*.

193 "We do know that today, introduced alien species. . . ." Novacek, M. J., and Cleland, E., 2001, "The current biodiversity extinction event: scenarios for mitigation and recovery," *Proceedings of the National Academy of Sciences* 98: 5466–70.

194 "The Nile perch. . . ." Stiassney, M. L. J., 2001, "Lake Victoria," in Novacek, *Biodiversity Crisis*, pp. 117–19.

194 "There is a long list of other harmful. . . ." Novacek and Cleland, "Current biodiversity extinction event."

194 "The whales were probably of Miocene age. . . ." Savage and Russell, *Mammalian Paleofaunas*.

18. A MAN AND HIS HORSE

196 "Ironically, the accident had an eerie resonance. . . ." Hatcher, *Bone Hunters in Patagonia*.

199 "Maps were opened and strewn about." The standard reference and geologic maps of the region are in Niemayer, H., 1975, "Geología de la region comprendida entre el Lago General Carrera y el río Chacabuco, Provincia de Aysen, Chile," thesis (Mem. Titulo), Department of Geology, University of Chile, Santiago; and Niemayer, H., Scarmeta, J., Fuenzalida, R., and Espinosa, W., 1984, "Hojas Peninsulada Taitao y Puerto Aysen Servicio Nacional de Geología y Mineria," *Carta Geologica de Chile*, nos. 60–61: 1–80.

201 ". . . what had been a very shallow lagoon. . . ." Niemayer et al., *Carta Geologica de Chile*.

209 "A faint trail wound through a plateau. . . ." Ibid.

19. PAMPA CASTILLO

217 ". . . spike-tailed glyptodont. . . ." Simpson, *Splendid Isolation*.

217 "These fossils were typical of Miocene series. . . ." Scott, W. B., 1912, "Mammalia of the Santa Cruz beds: Part 2: Toxodonta," *Reports of Princeton University Expeditions to Patagonia (1896–1899)*, no. 6: 111–238; Scott, W. B., 1928, "Mammalia of the Santa Cruz beds: Part 4: Astrapotheria," ibid., vol. 6: 301–42; Scott, W. B., 1930, "A partial skeleton of *Homalodotherium* from the Santa Cruz beds of Patagonia," *Memoirs of the Field Museum of Natural History* 50 (1): 1–39; Scott, W. B., 1937, "The Astrapotheria," *Proceedings of the American Philosophical Society* 77: 309–93; Simpson, G. G., 1948, "The beginning of the age of mammals in South America, Part I," *Bulletin of the American Museum of Natural History* 91: 1–232; Simpson, G. G., 1967, "The beginning of the age of mammals in South America, Part 2," *Bulletin of the American Museum of Natural History* 137: 1–260; Prothero, D. R., 1994, "Mammal evolution," in Prothero, D. R., and Schoch, R. M. (eds.), *Major Features of Vertebrate Evolution*, Short Courses in Paleontology, no. 7, Paleontological Society, Knoxville, TN; Croft, D., 1999, "Placentals: Endemic South American ungulates," *Encyclopedia of Paleontology*, Fitzroy Dearborn Publishers, Chicago, pp. 890–906.

220 ". . . near 'Pampa Castillo' . . . ." The work at this site has been recently reviewed in Flynn, J. J., Novacek, M. J., Dodson, E. D., Frassinetti, D., McKenna, M. C., Norell, M. A., Sears, E. S., Swisher III, C. C., Wyss, A. R., 2001, "A new fossil mammal assemblage from the Chilean Andes: Implications for geology, geochronology, and tectonics," *Journal of South American Earth Sciences* (in press).

222 ". . . glyptodont bones and the teeth and jaws of rodents. . . ." Ibid.

223 "Pampa Castillo was indeed at the ends of the earth. . . ." Ibid.

20. BONE HUNTERS IN PATAGONIA

226 ". . . teeth that belonged to a notoungulate." A useful summary of the South American ungulates is Croft, "Placentals: Endemic South American ungulates."

226 ". . . a big rhinolike toxodont. . . ." Ibid.

226 "The census of our fossil community. . . ." Ibid.

227 ". . . a tiny jaw and teeth of a rare, opossumlike marsupial." Patterson and Pascual, "Fossil mammal fauna."

229 "They ran out from the pack. . . ." Stehli, F. G., and Webb, S. D. (eds.), 1985, *The Great American Biotic Interchange*, Plenum Press, New York.

232 "As noted, the animals. . . ." Scott, "Mammalia of the Santa Cruz beds, Part 2," "Mammalia of the Santa Cruz beds, Part 4," "Partial Skeleton," and "Astrapotheria"; Simpson, "Beginning of the age of mammals"; Prothero, "Mammal evolution"; Croft, "Placentals."

233 ". . . the Andes were among the fastest-growing mountains in the world." Miller, *Continents in Collision*.

233 "*The New York Times* . . . 'Whale Fossils High in the Andes Show How Mountains Rose from Sea.' " *The New York Times,* March 12, 1987.

233 "A letter to the editor. . . ." Silver, J., 1987, "Darwin, too, saw the ocean in the Andes," *The New York Times* (letter to the editor), March 15, 1987.

21. ABOVE THE CLOUDS AND THE CONDORS

234 ". . . indicated Early Tertiary rocks. . . ." Klohn, C., 1957, "Estado actual del estudio geologica de la 'Formacion Porfiritica,' " *Review Minerales* 55: 1–12; Klohn, C., 1960, "Geología de la Cordillera de los Andes de Chile Central; provincias de Santiago, O'Higgins, Colchagua y Curico," *Publicación de Instituto del Investigación Geológica Boletin* 8: 1–95;

Charrier, R., 1981, "Mesozoic and Cenozoic stratigraphy of the central Argentinian-Chilean Andes (32°–35°) and chronology of their tectonic evolution," Zentralbl. für Geologie und Palaontologie 1 (3/4): 344–55.

234 ". . . magnificent dinosaur trackways. . . ." Lockley, M., 1995, "Track records," Natural History 104 (6): 46–51.

239 "Termas del Flaco and its dinosaur tracks. . . ." Klohn, "Estado actual."

240 ". . . it was like a Grand Canyon tilted on its side." Ibid.

240 "The top of the escarpment. . . ." Ibid.

241 ". . . marks that became impressions in horizontal beds. . . ." Lockley, "Track records."

242 "But the dominant and largest tracks. . . ." Ibid and Norell, Gaffney, and Dingus, Discovering Dinosaurs.

242 ". . . limestone sprinkled with specks of fish vertebra." Klohn, "Estado actual" and "Geología."

242 ". . . rocks full of extraordinary mammal skulls, jaws, and limb bones." The discovery was first disclosed in Novacek, M. J., Wyss, A., Frazinetti, D., and Salinas, P., 1989, "A new Eocene mammal fauna from the Andean Main Range," Journal of Vertebrate Paleontology, supplement to no. 3, 9: 34A.

243 "These were rather weird forms. . . ." Ibid.

243 "Oddly enough, only Mesozoic rock had been mapped. . . ." Klohn, "Estado actual" and "Geología"; Charrier, "Mesozoic and Cenozoic stratigraphy."

243 "The rocks of the mine were Cretaceous marine layers. . . ." Charrier, "Mesozoic and Cenozoic stratigraphy."

244 ". . . the bombing of the Presidential Palace. . . ." A comprehensive BBC review of Pinochet's rise to power and reign of terror can be found at http://news6.thdo.bbc.co.uk/hi/english/world/americas/newsid_63000/63821.stm.

246 ". . . Cretaceous Colimapu Formation." Klohn, "Estado actual" and "Geología"; Charrier, "Mesozoic and Cenozoic stratigraphy."

246 "So began a gloriously productive project. . . ." The relevant papers include Wyss, A. R., Norell, M. A., Flynn, J. J., Novacek, M. J., Charrier, R., McKenna, M. C., Swisher III, C. C., Frazinetti, D., and Jin, M., 1990, "A new Early Tertiary mammal fauna from central Chile: Implications for Andean stratigraphy and tectonics," Journal of Vertebrate Paleontology 10: 518–22; Wyss, A. R., Norell, M. A., Novacek, M. J., and Flynn, J. J., 1992, "New Early Tertiary localities from the Chilean Andes," Journal of Vertebrate Paleontology, supplement to no. 3, 12: 61A; Wyss, A. R., Flynn, J. J., Norell, M. A., Swisher III, C. C., Novacek, M. J., McKenna, M. C., and Charrier, R., 1993, "South America's earliest rodent and recognition of a new interval of mammalian evolution," Nature 365: 434–37; Wyss, A. R., Charrier, R., and Flynn, J. J., 1996, "Fossil mammals as a tool in Andean stratigraphy: dwindling evidence of Late Cretaceous volcanism in the South Central Main Range," PaleoBios 17 (2–4): 13–27.

247 ". . . well-preserved skulls, jaws, and skeletons of various mammals. . . ." Wyss, A. R., Flynn, J. J., Norell, M. A., Swisher III, C. C., Novacek, M. J., McKenna, M. C., and Charrier, R., 1994, "Paleogene mammals from the Andes of central Chile: A preliminary taxonomic, biostratigraphic, and geochronologic assessment," American Museum of Novitates, no. 3098: 1–31.

247 "The dark stories of the survivors. . . ." Read, P. P., Alive!, Avon, New York, 1979.

248 "The discoveries above the Tinguiririca Valley. . . ." Wyss et al., "A new early Tertiary mammal fauna"; Wyss et al., "New Early Tertiary localities"; Charrier, R., Wyss, A. R., Flynn, J. J., Swisher III, C. C., Norell, M., Zapatta, F., McKenna, M. C., and Novacek, M. J., 1996, "New evidence for late Mesozoic–early Cenozoic evolution of the Chilean Andes in the upper Tinguiririca Valley (35°S), Central Chile," Journal of South American Geology 9 (5/6): 393–422.

248 "Soon it became apparent. . . ." Wyss et al., "Paleogene mammals"; Wyss, Charrier, and Flynn, "Fossil mammals as a tool in Andean stratigraphy."

248 ". . . a beautiful jaw of a primitive New World monkey." Flynn, J. J., Wyss, A. R., Charrier, R., Swisher III, C. C., 1995, "An early Miocene anthropoid skull from the Chilean Andes," *Nature* 373: 603–07.

249 ". . . Renaldo Charrier. . . ." Charrier et al., "New evidence."

22. THE LAND OF SHEBA

253 "Yemen was a strange land of contrasts. . . ." A similar impression from one who traveled the country that same season is Hansen, Eric, 1991, *Motoring with Mohammed*, Houghton Mifflin, Boston.

254 ". . . Yemen offered dramatic possibilities for fossils." This was based on both discoveries of mammal fossils elsewhere on the Arabian Peninsula as well as compelling arguments for the potential of finding more. Madden, C. T., Schmidt, D. L., and Whitmore, F. C., 1983, "*Mastritherium* (Artiodactyla, Anthractheriidae) from Wadi Sabya, southwestern Saudi Arabia—An earliest Miocene age for continental rift-valley volcanic deposits of the Red Sea margin," Saudi Arabian Deputy Ministry for Mineral Resources Open-file Report USGS-OF-03-61 (Interagency Report IR-561), pp. 1–24; Whybrow, Peter, 1984, "Geological and faunal evidence from Arabia for mammal 'migrations' between Asia and Africa during the Miocene," *Courier Forschunginstitut Senckenberg* 69: 169–98; Whybrow, Peter, 2000, "Arabia Felix, fossilised fruits and the price of frogs," in Whybrow, Peter (ed.), *Travels with Fossil Hunters*, Natural History Museum Publishing, Cambridge University Press in association with the Natural History Museum, London, pp. 196–211.

254 ". . . the Amran Series. . . ." Geukens, F., 1966, *Geology of the Arabian Peninsula, Yemen*, U.S. Geological Survey Professional Paper 560-B, including geologic map, scale 1:1,000,000; Grolier, M. J., Domenico, J. A., Donato, M., Tibbitts, G. C., Jr., Overstreet, W. C., and Ibrahim, M. M., 1977, "Data from geologic investigations in the Yemen Arab Republic during 1976," *U.S Geological Survey Report Yemen Arab Republic Investigations* (IR) Y-12.

255 ". . . Cretaceous Tawilah sandstones. . . ." Grolier, M. J., and Overstreet, W. C., 1978, *Geologic Map of the Yemen Arab Republic (San'a')*, U.S. Geological Survey Miscellaneous Investigations Series I-1143-B, scale 1:500,000.

255 ". . . Rift Zone. . . ." Ibid.

255 "As Peter Whybrow . . . argued. . . ." Whybrow, "Geological and faunal evidence."

263 ". . . the geologic maps. . . ." Geukens, *Geology of the Arabian Peninsula*; Grolier et al., "Data from geologic investigations"; Grolier and Overstreet, *Geologic Map*.

263 ". . . Carsten Niebuhr. . . ." Niebuhr, C., 1994, *Travels through Arabia and Other Countries of the East*, translated by Robert Heron, Garnet, Reading, PA.

265 ". . . Pleistocene rocks near a road. . . ." Geukens, *Geology of the Arabian Peninsula*; Grolier et al., "Data from geologic investigations"; Grolier and Overstreet, *Geologic Map*.

266 ". . . Jurassic-aged fish. . . ." Carroll, *Vertebrate Paleontology*.

274 ". . . some fine fish skeletons." Ibid.

274 ". . . other evidence of Mesozoic life. . . ." Ibid.

276 "A few eroded monuments of this past glory. . . ." Interesting information on Mareb's ancient structures, notably the original dam dating from about 1000 B.C., can be found at *http://www.yementimes.com/98/iss52/lastpage.htm*.

23. TO THE FOOT OF THE FLAMING CLIFFS

278 ". . . ray-finned fishes (leptolepids). . . ." Carroll, *Vertebrate Paleontology*, pp. 113–14.

279 ". . . the legendary Roy Chapman Andrews." Andrews, R. C., 1932, *The New Conquest*

of Central Asia, vol. 1 of *Natural History of Central Asia*, American Museum of Natural History, New York. A recent biography of Andrews is Gallenkamp, C., 2001, *Dragon Hunter*, Viking Penguin, New York.

279 "It was also for the American Museum. . . ." Preston, *Dinosaurs in the Attic.*

280 "Then there were those dinosaurs. . . ." Colbert, *Men and Dinosaurs.*

280 ". . . all great men of paleontology. . . ." Ibid.

280 "According to several highly respected experts like William Diller Matthew. . . ." Matthew, W. D., 1915, "Climate and Evolution," *Annals of the New York Academy of Sciences* 24: 171–318.

280 "This appealed to Osborn for both scientific and less distinguished motives." Gallenkamp, *Dragon Hunter.*

281 "The Gobi was not trackless. . . ." Haslund, H., 1934, *In Secret Mongolia*, Mystic Travelers Series, Adventure Unlimited Press (first printing 1995).

281 "The first expedition set out from Beijing in 1922. . . ." Andrews, *New Conquest*; Novacek, *Dinosaurs of the Flaming Cliffs*; Gallenkamp, *Dragon Hunter.*

281 " 'Well, Roy, we've done it!' " Gallenkamp, *Dragon Hunter.*

281 ". . . John B. Shackelford. . . ." Ibid.

282 "The 1923 Central Asiatic Expedition. . . ." Ibid.

282 ". . . which Andrews called the Flaming Cliffs. . . ." Ibid.

283 ". . . scores of the beak-headed dinosaur *Protoceratops*. . . ." Ibid.

283 "One triumphant field season followed another." Ibid.

283 ". . . skulls of small mammals. . . ." Ibid.

284 "This succession of paleontological victories had a downside." Ibid.

285 ". . . a big *Tyrannosaurus* named Sue. . . ." Fiffer, S., 2000, *Tyrannosaurus Sue: The Extraordinary Saga of the Largest, Most Fought Over T. Rex Ever Found*, Freeman, New York.

285 ". . . Choibalsan's purge of Buddhists. . . ." Storey, R., 1993, *Mongolia, a Travel Survival Kit*, Lonely Planet Publications, Hawthorn, Australia.

285 "It was not until 1946. . . ." Efremov, I. A., 1956, *The Way of the Winds*, All-Union Pedagogical-Study Publisher, Ministry of Reserve Workers, Moscow (in Russian).

286 "The findings included a magnificent *Tarbosaurus*. . . ." Ibid.

286 "Joint Polish-Mongolian expeditions. . . ." Kielan-Jaworowska, Z., 1969, *Hunting for Dinosaurs*, Maple Press, York, PA; Lavas, J. R., 1993, *Dragons from the Dunes*, Academy Interprint, Auckland, New Zealand; Kielan-Jaworowska, Z., 1975, "Late Cretaceous dinosaurs and mammals from the Gobi desert," *American Scientist* 63: 150–59; Gradzinski, R., Kielan-Jaworowska, Z., and Maryanska, T., 1977, "Upper Cretaceous Djadokhta, Barun Goyot and Nemegt formations of Mongolia, including remarks on previous subdivisions," *Acta Geologica Polonica* 27 (3): 281–318.

286 ". . . some precious new mammals like *Kennalestes*." Gradzinski, Kielan-Jaworowska, and Maryanska, "Upper Cretaceous."

286 "Here, the paleontologists found something remarkable. . . ." Ibid.

286 ". . . Canadians conducted a series of important and productive collaborations. . . ." Currie, P. J. (ed.), 1993, "Results from the Sino-Canadian dinosaur project," *Canadian Journal of Earth Sciences* 30 (10–11): 1997–2272.

289 "I thought of Andrews. . . ." Andrews, *New Conquest.*

291 ". . . the famous fighting pair. . . ." Kielan-Jaworowska, *Hunting for Dinosaurs.*

292 ". . . a partial skeleton of a large lizard." Norell, M. A., McKenna, M. C., and Novacek, M. J., 1992, "*Estesia mongoliensis*, a new fossil varanoid from the Late Cretaceous Barun Goyot Formation of Mongolia," *American Museum Novitates*, no. 3045: 1–24.

292 ". . . Mongolian Academy–American Museum Paleontological Expedition." Novacek, M. J., Norell, M. A., McKenna, M. C., and Clark, J., 1994, "Fossils of the Flaming Cliffs," *Scientific American* 271 (6): 60–69; and Novacek, *Dinosaurs of the Flaming Cliffs.*

24. THE CRETACEOUS CORNUCOPIA

293 "And farther still. . . ." Andrews, *New Conquest.*

294 ". . . another *Oviraptor* skeleton" Smith, D., 1992, "The type specimen of *Oviraptor philoceratops,* a theropod dinosaur from the Upper Cretaceous of Mongolia," *Neus Jahrsbuch Geol. Palaont. Abh.,* pp. 365–88; Norell, M. A., Clark, J. M., Chiappe, L. M., and Dashzeveg, D. M., 1995, "A nesting dinosaur," *Nature* 378: 774–76.

294 "We were all there, at this place called Ukhaa Tolgod. . . ." Dashzeveg, D., Novacek, M. J., Norell, M. A., Clark, J. M., Chiappe, L. M., Davidson, A., McKenna, M. C., Dingus, L., Swisher, C., and Perle, A., 1995, "Extraordinary preservation in a new vertebrate assemblage from the Late Cretaceous of Mongolia," *Nature* 374: 446–49; Novacek, *Dinosaurs of the Flaming Cliffs.*

294 ". . . the important mammal *Deltatheridium.*" Rougier, G. W., Wible, J. R., and Novacek, M. J., 1998, "Implications of *Deltatheridium* specimens for early marsupial history," *Nature* 396: 459–63.

295 ". . . the big lizard *Estesia*. . . ." Norell, McKenna, and Novacek, "*Estesia mongoliensis.*"

296 "In 1226, the Year of the Dog, Temujin, alias Genghis Khan. . . ." Bazargur, D., and Enkhbayer, D., 1996, *Chinggis Khaan: Historic-Geographic Atlas,* Ulan Bator, Mongolia, pp. 46–47.

296 "Some have even proposed that *Protoceratops* skeletons were the inspiration for griffins. . . ." Mayor, A., 2000, *The First Fossil Hunters,* Princeton University Press, Princeton, NJ.

296 "Some spots in the Gobi are surprisingly thick with skeletons." Novacek, *Dinosaurs of the Flaming Cliffs.*

298 ". . . the flightless bird *Mononykus*. . . ." Perle, A., Chiappe, L. M., Barsbold, R., Clark, J. M., and Norell, M. A., 1994, "Skeletal morphology of the *Mononykus olecranus* from the late Cretaceous of Mongolia," *American Museum Novitates,* no. 3105: 1–29.

298 ". . . the hopping mammal *Zalambdalestes*. . . ." Novacek, "A pocketful of fossils."

299 "I have elsewhere related the trials and adventures of the Gobi expeditions. . . ." Novacek, *Dinosaurs of the Flaming Cliffs.*

299 ". . . he had found the best fossil ever." Norell, M. A., Clark, J. M., Dashzeveg, M. J., and Novacek, M. J., 1994, "A therapod dinosaur embryo and the affinities of the Flaming Cliffs dinosaur eggs," *Science* 266: 779–82.

301 "The nesting oviraptorid. . . ." Norell et al., "A nesting dinosaur"; Clark, J. M., Norell, M. A., and Chiappe, L. M., 1999, "An oviraptorid skeleton from the late Cretaceous of Ukhaa Tolgod, Mongolia, preserved in an avianlike brooding position over an oviraptorid nest," *American Museum Novitates,* no. 3265: 1–36.

302 "This specimen was retrieved from Bayan Mandahu. . . ." Jerzykiewicz, T., Currie, P. J., Eberth, D. A., Johnston, P. A., Koster, E. H., and Zheng, J.-J., 1993, "Djadokhta Formation correlative strata in Chinese Inner Mongolia: an overview of the stratigraphy, sedimentary geology, and paleontology and comparisons with the type locality of the pre-Altai Gobi," *Canadian Journal of Earth Sciences* 30 (10–11): 2180–95.

302 "In 1995 we found two of these complete skeletons. . . ." This discovery was featured in Webster, D., 1996, "Dinosaurs of the Gobi: unearthing a fossil trove," *National Geographic* 190: 73–89.

302 "One I uncovered in 1993 was a small troodontidlike. . . ." Norell, M. A., Makovicky, P. J., and Clark, J. M., 2000, "A new troodontid theropod from Ukhaa Tolgod, Mongolia," *Journal of Vertebrate Paleontology* 20 (1): 7–11.

302 "The biggest dinosaurs at Ukhaa are the ankylosaurs. . . ." Novacek, *Dinosaurs of the Flaming Cliffs.*

303 "When extracting a big ankylosaur. . . ." Rougier, G. W., Novacek, M. J., and Dashzeveg, D., 1997, "A new multituberculate from the Late Cretaceous locality Ukhaa Tolgod,

Mongolia. Considerations of multituberculates' interrelationships," *American Museum Novitates*, no. 3191: 1–26; Keqin, Gao, and Norell, M. A., 1998, "Taxonomic revision of *Carusia* (Reptilia Squamata) from the Late Crtetaceous of the Gobi Desert and phylogenetic relationships of Anguimorphan lizards," *American Museum Novitates*, no. 3230: 1–51.

303 "A new form called *Ukhaatherium*. . . ." Novacek, M. J., Rougier, G. W., Wible, J. R., McKenna, M. C., Dashzeveg, D., and Horovitz, I., 1997, "Epipubic bones in eutherian mammals from the Late Cretaceous in Mongolia," *Nature* 389: 483–86.

304 "It has been suspected that the epipubics. . . ." Ibid.

305 "In pondering this problem. . . ." Ibid.

305 "In a paper published in 1997. . . ." Ibid.

305 ". . . the sharp-toothed deltatheridians. . . ." Rougier, Wible, and Novacek, "Implications of *Deltatheridium*."

306 ". . . a newborn marsupial is so poorly developed. . . ." Ibid.

306 "*Deltatheridium* had tooth, skull, and skeletal features. . . ." Ibid.

306 "Several aspects of the fossil concentrations gave us some clues to the answers." Novacek, *Dinosaurs of the Flaming Cliffs*.

307 "Since the expeditions of the American Museum in the 1920s. . . ." Ibid.

307 ". . . he resolutely paced the outcrops for clues." Results of this investigation were published in Loope, D. B., Dingus, L., Swisher III, C. C., and Minjin, C., 1998, "Life and death in a Cretaceous dune field," *Geology* 26: 27–30. The analysis was later reviewed in Dingus, L., and Loope, D., 2000, "Death in the dunes," *Natural History* 109 (6): 50–55.

308 ". . . huge rainstorms caused mudflows. . . ." Dingus and Loope, "Death in the dunes."

310 "The harsh oscillations of drought and flood. . . ." Ibid.

310 "When a *Tyrannosaurus* sells at auction. . . ." Fiffer, *Tyrannosaurus Sue*.

312 "Mongolia is indeed a country transformed. . . ." Novacek, *Dinosaurs of the Flaming Cliffs*; Man, J., 1997, *Gobi: Tracking the Desert*, Weidenfeld & Nicolson, London.

25. LAST CHANCE CANYON

314 "The rocks of Cerro Condor are tawny. . . ." A general recent review of dinosaur distribution and localities in South America with a summary of literature is Novas, F. E., 1997, "South American dinosaurs," in Currie, P. J., and Padian, K. (eds.), *Encyclopedia of Dinosaurs*, Academic Press, San Diego, pp. 678–89.

315 ". . . the big predator *Carnotaurus*. . . ." Ibid.

315 ". . . his find of *Amargasaurus*. . . ." Ibid.

316 ". . . Australia and South America were broadly connected through Antarctica." Schopf, *Major Events in the History of Life*; Miller, *Continents in Collision*.

318 "The Cretaceous extinction event that erased the nonbird dinosaurs. . . ." Archibald, *Dinosaur Extinction*.

318 "But the golden age of exploration exploited that unfamiliar terrain." Preston, *Dinosaurs in the Attic*.

319 ". . . a terrified man in a lawn chair. . . ." The intention of the Darwin Awards is to celebrate Darwin's theory of evolution by commemorating those who have contributed to improving our gene pool by eliminating themselves in acts of stupidity. The lawn chair pilot survived and thus did not quite qualify, but the act deserves credit nonetheless. Northcutt, W., 2000, *The Darwin Awards: Evolution in Action*. The popular Web site is *http://www.darwinawards.com*.

319 ". . . Ernest Shackleton's 1914 attempt to cross the Antarctic continent. . . ." Alexander, C., 1998, *The Endurance: Shackleton's Antarctic Expedition*, Knopf, New York.

320 ". . . the world had experienced an infusion of unprecedented scientific knowledge."
     Pyne, S. J., 1998, *The Ice: A Journey into Antarctica*, University of Washington Press, Seattle.
320 "A recent scientific paper has shown that there may be a ten-million-year lag. . . ."
     Kirchner and Weil, "Delayed biological recovery."
321 "Many scientists refer to this biodiversity crisis as the sixth mass extinction." Eldredge,
     N., 2001, "Evolution, extinction, and humanity's place in nature," in Novacek, *Biodiver-
     sity Crisis*, pp. 76–80.
321 ". . . document the vanishing biota. . . ." Ibid.
321 "Such momentous issues are admittedly distracting. . . ." Myers, N., 2001, "What's this
     biodiversity and what's it done for us today?" in Novacek, *Biodiversity Crisis*, pp. 21–25.

Selected Reading List

■ ■ ■

Alt, D., and Hyndman, D. W., 1986. *Roadside Geology of Montana*. Mountain Press Publishing, Missoula, MT.

Andrews, R. C., 1932. *The New Conquest of Central Asia*. Vol. 1 of *Natural History of Central Asia*. American Museum of Natural History, New York.

Archibald, J. D., 1966. *Dinosaur Extinction at the End of an Era: What the Fossils Say*. Columbia University Press, New York.

Baars, D. L., 2000. *The Colorado Plateau: A Geologic History*. University of New Mexico Press, Albuquerque.

Berry, W. B. N., 1968. *Growth of a Prehistoric Time Scale*. Freeman, San Francisco.

Carroll, R. L., 1988. *Vertebrate Paleontology and Evolution*. Freeman, New York.

Chronic, H., 1987. *Roadside Geology of New Mexico*. Mountain Press Publishing, Missoula, MT.

Cloud, P. (ed.), 1970. *Adventures in Earth History*. Freeman, San Francisco.

Colbert, E. H., 1968. *Men and Dinosaurs*. Dutton, New York.

————, 1995. *The Little Dinosaurs of Ghost Ranch*. Columbia University Press, New York.

Cox, A. (ed.), 1973. *Plate Tectonics and Geomagnetic Reversals*. Freeman, San Francisco.

Currie, P. J. and Padian, K. (eds), 1997. *Encyclopedia of Dinosaurs*. Academic Press, San Diego.

Dalrymple, G. B., 1991. *The Age of the Earth*. Stanford University Press, Stanford, CA.

Dashzeveg, D., Novacek, M. J., Norell, M. A., Clark, J. M., Chiappe, L. M., Davidson, A., McKenna, M. C., Dingus, L., Swisher, C., and Perle, A., 1995. "Extraordinary preservation in a new vertebrate assemblage from the Late Cretaceous of Mongolia." *Nature* 374: 446–49.

Dingus, L., and Loope, D., 2000. "Death in the dunes." *Natural History* 109 (6): 50–55.

Dingus, L., and Rowe, T., 1998. *The Mistaken Extinction: Dinosaur Evolution and the Origin of Birds.* Freeman, New York.

Eldredge, N., 1985. *Time Frames.* Simon and Schuster, New York.

Erwin, D. H., 1993. *The Great Paleozoic Crisis.* Columbia University Press, New York.

Gee, H., 1999. *In Search of Deep Time; Beyond the Fossil Record to a New History of Life.* Free Press, Simon and Schuster, New York.

———(ed.), 2000. *Shaking the Tree: Readings from Nature in the History of Life.* University of Chicago Press, Chicago.

Glen, W., 1990. "What killed the dinosaurs?" *American Scientist* 78 (4): 354–70.

Grolier, M. J., and Overstreet, W. C., 1978. *Geologic Map of the Yemen Arab Republic (San'a').* U.S. Geologic Survey Miscellaneous Investigations Series I-1143-B, scale 1:500,000.

Harland, W. B., Armstrong, R. L., Cox, A. V., Craig, L. E., Smith, A. G., and Smith, D. G., 1990. *A Geologic Time Scale, 1989 Edition.* Cambridge University Press, Cambridge, England.

Harris, J. M., and Jefferson, G. T. (eds.), 1985. *Rancho La Brea: Treasures of the Tar Pits.* Natural History Museum of Los Angeles County.

Hatcher, J. B., 1985. *Bone Hunters in Patagonia, Narrative of the Expedition.* Ox Bow Press, Woodbridge, CT. Reprinted from Hatcher, J. B., 1903. *Princeton University Expedition to Patagonia, 1896–1899.* Princeton University.

Kielan-Jaworowska, Z., 1969. *Hunting for Dinosaurs.* Maple Press, York, PA.

———, 1975. "Late Cretaceous dinosaurs and mammals from the Gobi Desert." *American Scientist* 63: 150–59.

Lageson, D. R., and Spearing, D. R., 1988. *Roadside Geology of Wyoming.* Mountain Press Publishing, Missoula, MT.

Mathez, E. A. (ed.), 2001. *Earth: Inside and Out.* American Museum of Natural History and New Press, New York. Distributed by Norton.

McPhee, J., 1980. *Basin and Range.* Farrar, Straus & Giroux, New York.

———, 1986. *Rising from the Plains.* Farrar, Straus & Giroux, New York.

Miller, R., 1983. *Continents in Collision.* Time-Life Books, Alexandria, VA.

Milner, R., 1993. *The Encyclopedia of Evolution.* Henry Holt, New York.

Nelson, G., and Platnick, N. I., 1981. *Systematics and the Biogeography—Cladistics and Vicariance.* Columbia University Press, New York.

Niebuhr, C., 1994. *Travels through Arabia and Other Countries of the East.* Translated by Robert Heron. Garnet, Reading, PA.

Norell, M. A., 1995. "Origins of the feathered nest." *Natural History* 104 (6): 58–61.

Norell, M. A., Clark, J. M., Dashzeveg, M. J., and Novacek, M. J., 1994. "A theropod dinosaur embryo and the affinities of the Flaming Cliffs dinosaur eggs." *Science* 266: 779–82.

Norell, M. A., Gaffney, E. S., and Dingus, L., 1995. *Discovering Dinosaurs.* Knopf, New York.

Novacek, M. J., 1994. "A pocketful of fossils." *Natural History* 103 (4): 40–43.

———, 1996. *Dinosaurs of the Flaming Cliffs.* Anchor/Doubleday, New York.

———, 1999. "100 million years of land vertebrate evolution: the Cretaceous-Early Tertiary Transition." *Annals of the Missouri Botanical Gardens* 86: 230–58.

———(ed.), 2001. *The Biodiversity Crisis: Losing What Counts.* American Museum of Natural History and New Press, New York. Distributed by Norton.

Novacek, M. J., Flynn, J. J., Ferrusquia-Villafranca, I., and Cipolletti, R. M., 1987. "An early Eocene (Wasatchian) mammal fauna from Baja California." *National Geographic Research* 3: 376–88.

Novacek, M. J., Norell, M. A., McKenna, M. C., and Clark, J., 1994. "Fossils of the Flaming Cliffs." *Scientific American* 271 (6): 60–69.

Novacek, M. J., and Wheeler, Q. D. (eds.), 1992. *Extinction and Phylogeny.* Columbia University Press, New York.

Ostrom, J. H., 1974. *"Archaeopteryx* and the origin of flight." *Quarterly Review of Biology* 49: 27–47.

Preston, D. J., 1986. *Dinosaurs in the Attic: An Excursion into the American Museum of Natural History.* St. Martin's Press, New York.

Simpson, G. G., 1934. *Attending Marvels: A Patagonian Journal.* University of Chicago Press, Chicago.

———, 1980. *Splendid Isolation: The Curious History of South American Mammals.* Yale University Press, New Haven, CT.

Stehli, F. G., and Webb, S. D. (eds.), 1985. *The Great American Biotic Interchange.* Plenum Press, New York.

Szalay, F. S., Novacek, M. J., and McKenna, M. C. (eds.), 1993. *Mammal Phylogeny (Vol. 1): Mesozoic Differentiation, Multituberculates, Monotremes, Early Therians, and Marsupials.* Springer Verlag, New York.

———, 1993. *Mammal Phylogeny (Vol. 2): Placentals.* Springer Verlag, New York.

Tattersall, I., 1995. *The Fossil Trail.* Oxford University Press, Oxford, England.

Webster, D., 1996. "Dinosaurs of the Gobi: unearthing a fossil trove." *National Geographic* 190: 73–89.

Wyss, A. R., Flynn, J. J., Norell, M. A., Swisher III, C. C. R., Novacek, M. J., McKenna, M. C., and Charrier, R., 1993. "South America's earliest rodent and recognition of a new interval of mammalian evolution." *Nature* 365: 434–37.

Index

. . .

Illustration Credits

Illustrations on pages 6, 20, 40, 43, 68, 109, 110, 125, 200, 246, 251, 283, 290, 297, 299, 304, 306, 309, 316, and 317, drawn by Ed Heck.

Illustrations on pages 9, 68, 115, 130, 133, 134, 138, 142, 147, 149, 153, 160, 247, 282, and 284, courtesy American Museum of Natural History (AMNH) publications and archives.

Page 14, *Turritella*, drawing by Jerry Burgess, modified.

Page 17, "Panorama of Grand Canyon," courtesy United States Geological Survey.

Page 18, "A geologic section of the Grand Canyon," and page 191, "Ancient mammals of South America," from *Stratigraphy and Life History* by M. Kay and E. H. Colbert, copyright © 1965 by M. Kay and E. H. Colbert. This material is used by permission of John Wiley & Sons, Inc.

Page 27, "A trilobite," from *Time Frames* by Niles Eldredge, copyright © by Niles Eldredge. Reprinted with the permission of Simon & Schuster.

Page 37, "*Dimetrodon*," "*Edaphosaurus*," page 63, "Skulls of pelycosaurs," and page 132, "*Stegosaurus*," from *Osteology of the Reptiles* by A. S. Romer, copyright © 1956 by The University of Chicago Press. Reprinted with the permission of The University of Chicago Press.

Page 38, "*Diploceraspis*," courtesy of the Museum of Comparative Zoology, Harvard University.

Page 43, "The major tectonic plates," from *The Mistaken Extinction: Dinosaur Evolution and the Origin of Birds* by Lowell Dingus and Timothy Rowe, copyright © 1998 by Lowell Dingus and Timothy Rowe. Reprinted by permission of Henry Holt and Company, LLC.

Page 71, "*Thrynaxodon*," from "The postcranial skeleton of African cynodants" by F. A. Jenkins, Jr., in *Bulletin of the Peabody Museum of Natural History*, Vol. 36, 1971. Reprinted with the permission of The Peabody Museum of Natural History.

Page 71, "*Kannemeyeria*," from "A dicynodont reptile reconstructed," in *Proceedings Zoological Society of London* 1924, copyright © by H. S. Pearson. Reprinted with the permission of Cambridge University Press.

Page 77, "Upper jaw and teeth of *Desmatodon*," and page 92, "Disarticulated skeleton of *Trihecaton*," courtesy Dr. Peter P. Vaughn.

Page 85, "*Tsejaia*," from "The morphology and phylogenetic relationships of the Lower Permian tetrapod *Tseajaia campi* Vaughn (Amphibia: Seymouriamorpha)," copyright © 1972 by J. L. Moss. Reprinted with the permission of the *University of California Publications in Geological Science*, Vol. 98.

Page 86, "*Seymouria*," from *American Permian Vertebrates* by S. W. Williston, copyright © 1911 S. W. Williston. Reprinted with the permission of The University of Chicago Press.

Page 99, "*Canis diris,*" "*Smilodon californicus,*" page 156, "Teeth of Cretaceous and Cenozoic sharks," page 190, "Skull of *Thylacosmilus,*" and page 216, "A glyptodont," from *Vertebrate Paleontology* by A. S. Romer, copyright © 1966 by The University of Chicago. Reprinted with the permission of The University of Chicago Press.

Page 108, "Lower jaws of marsupials and placentals," from "An ontogenetic assessment of dental homologies in therian mammals" by W. P. Luckett, in *Mammal Phylogeny (Vol. 1): Mesozoic Differentiation, Multituberculates, Monotremes, Early Therians, and Marsupials,* edited by F. S. Szalay, M. J. Novacek, and M. C. McKenna, 1993. Reprinted with the permission of Springer-Verlag.

Page 118, "First lower molar and last lower premolar of *Batonoides,*" drawing by M. Novacek from "Insectivora and Proteutheria of the later Eocene (Uintan) of San Diego County, California," by M. J. Novacek, 1976. *Natural History Museum of Los Angeles County Contributions in Science,* No. 283.

Page 143, "Skull of *Triceratops,*" illustration by Donna Sloan, Royal Tyrrell Museum of Paleontology, from "Neoceratopsia" by P. Dodson and P. J. Currie, in *The Dinosauria,* edited by D. B. Weishampel, P. Dodson, and H. Osmolska, University of California Press, Berkeley, 1990.

Page 150, "*Pteranodon,*" courtesy of the Connecticut Academy of Arts and Sciences.

Page 173, "Our fine locality, Lomas Las Tetas de Cabra," photograph by M. Novacek.

Page 176, "Lower jaw and teeth of *Wyolestes,*" photograph by Chester Tarka, AMNH.

Page 226, "*Nesodon,*" Scott, W. B., 1912. "Mammalia of the Santa Cruz beds: Part 2: Toxodonta," *Reports of Princeton University Expeditions to Patagonia (1896–1899),* No. 6.

Page 260, "A Yemen house," photograph by M. Novacek.

Page 291, "Skull of *Estesia,*" photograph by Chester Tarka, AMNH.

Page 300, "An oviraptorid embryo," and page 301, "A nesting oviraptorid," drawings by Mick Ellison, courtesy American Museum of Natural History (AMNH).

Page 315, "Skull of *Carnotaurus,*" courtesy of Dr. Jose F. Bonaparte, Argentino de Ciencias Naturales.